Francis William Galpin, Charles Candler

An Account of the Flowering Plants, Ferns and Allies of Harleston

With a Sketch of the Geology, Climate, and Natural Characteristics of...

Francis William Galpin, Charles Candler

An Account of the Flowering Plants, Ferns and Allies of Harleston
With a Sketch of the Geology, Climate, and Natural Characteristics of...

ISBN/EAN: 9783337014070

Printed in Europe, USA, Canada, Australia, Japan

Cover: Foto ©berggeist007 / pixelio.de

More available books at **www.hansebooks.com**

An Account

OF THE

Flowering Plants

FERNS AND ALLIES

OF

HARLESTON.

With a Sketch of the Geology, Climate, and Natural Characteristics of the Neighbourhood.

COMPILED AND EDITED BY

THE REV. FRANCIS WILLIAM GALPIN,

M.A., F.L.S.,

TRINITY COLLEGE, CAMBRIDGE; FORMERLY CURATE OF REDENHALL
WITH HARLESTON AND WORTWELL, NORFOLK.

TO WHICH ARE ADDED

OBSERVATIONS ON THE

BIRDS OF THE DISTRICT

BY

CHARLES CANDLER.

LONDON: BARTLETT & CO., 10, PATERNOSTER SQUARE
HARLESTON: R. R. CANN.

1888.

TO MY

FRIENDS AND FELLOW-WORKERS,

THE MEMBERS OF

THE HARLESTON BOTANICAL CLUB.

———

"*A plant gathered in some delightful spot is more dear to memory than even a portrait.*"—SIR J. SMITH.

PREFACE.

THE following attempt to describe the Flowering Plants and Ferns of Harleston is the work, not of one, but of many. It owes its existence to the observations of valued friends found in my first curacy, whose kindred tastes formed an immediate bond of association between us, and became the source of much pleasant intercourse and, I trust, of mutual benefit. To the records of the Botanical Club have been added those of another personal friend, the late Rev. E. A. Holmes, Rector of St. Margaret's, whose kindly interest and scientific experience were ever extended to those who were endeavouring to trace the wondrous paths of nature which he had himself known for more than half a century.

Having, however, undertaken the duty of editing the result of our combined labours, I have probably made myself responsible for the truthfulness of the statements given. I trust it will be sufficient to say that of the six hundred and seventy-six species recorded by the Club, all except twenty-six have passed under my own notice; that of the remaining one hundred and fifteen species reported, more than half are confirmed by the authority of Mr. Holmes; and all records resting on other authority have been carefully examined, the plants being considered extinct if recent observation would probably have confirmed their existence but has failed to do so—marked as introduced if the locality favoured that supposition—rejected as unwarranted if upon correspondence no specimens or details were forthcoming. It asks a little self-denial to refuse a reported discovery which has been long desired and would augment the district list, yet cannot be satisfactorily verified; such self-denial, however, has not been wanting.

I am sorry that at the time of my departure from the neighbourhood the Cryptogamic Orders were practically untouched: from observation and report I have reason to believe that, were an enthusiast to arise, he would find an ample field for his energies, especially among the mushrooms and toadstools of a most fungiferous locality.

In other branches of Physical History, however, the neighbourhood has not been neglected; the Geology has received the careful attention of my friend and fellow-worker Mr. Charles Candler, who has supplied many details for the present general sketch, though he is unwilling to commit himself to any theories therein contained; to him also the reader is indebted for the able and interesting observations on the Birds. The Butterflies and Moths have been partially recorded by the late Mr. James Muskett, by Mr. Candler and his brothers, and are now being examined more completely and critically by the Rev. C. T. Cruttwell, Rector of Denton, who has also commenced a record of the Beetles and Flies with a view to publication at a future date. The Conchology has received the attention of Mr. Edmund Candler, but as his list is at present imperfect, it has been thought advisable to defer it.

The lack of works of reference, which a country curacy necessitated, has been amply supplied by the Libraries of the British Museum and the Linnean Society, in which many of the following details have been written and revised. But the duties of a large London parish, "in the fields" by courtesy only, have given little leisure; I must, therefore, apologise for the apparent brevity of certain portions of the work, and for any editorial inaccuracies which have found place in it, notwithstanding the careful scrutiny of my brother and Mr. Walter Cordwell, who have kindly looked through the proof-sheets.

As I draw my pleasant task to a close, I cannot but take for my own the quaint words in which a kinsman of the last century, John Galpine, has concluded his efforts in the same branch of Natural History—with him "disclaiming any merit whatever on my part, further than endeavouring to promote the cultivation of one of the most innocent, rational, and useful accomplishments."

F. W. G.

LONDON.
S. Giles in the Fields,
Christmas, 1887.

CONTENTS.

I.—INTRODUCTION:

    A. Local Features.
- (i.) Geology.
- (ii.) The Stone Age.
- (iii.) Climate.

    B. Botanical Research.
- (i.) Past Observers.
- (ii.) The Harleston Botanical Club.

    C. Floral Characteristics.
- (i.) The British Flora compared.
- (ii.) Flora of Norfolk and Suffolk compared.
- (iii.) Traces of a Maritime Flora.

II.—BOTANICAL BOOKS AND COLLECTIONS.

III.—AUTHORITIES, ABBREVIATIONS AND SIGNS.

IV.—A LIST OF THE PLANTS.

V.—OBSERVATIONS ON THE BIRDS.

VI.—INDICES.

I.—INTRODUCTION.

The market-town of HARLESTON, the centre of the Botanical District described in the following pages, is situated in lat. 52° 24′ N., and long. 1° 18′ E., about nineteen miles due south of the city of Norwich and eighteen miles due west of the coast-line, on the southern border of the county of Norfolk. It is supposed to owe its origin to Herolf the Dane, who, at the beginning of the eleventh century, obtained a grant of the narrow strip of land on which the middle row of houses is now built. The large stone near the Reading-room, popularly known as Herolf's Stone, is only one of the many boulders incidental to the geological nature of the district, though its opportune presence may have suggested its use as a landmark. In the area embraced by the six-mile circle, which is taken as a convenient limit of observation, there are fifty-five village parishes and parts of parishes, thirty-two being in Norfolk and twenty-three in Suffolk, with a total population of over 23,000 inhabitants, of which about 1,500 are concentrated in or near the central town.

A.—LOCAL FEATURES.

The country around Harleston is undulating, with occasional woodlands, and wholly devoted to agriculture. The highest points attained above the sea are—in Norfolk, on the lands between Shelton and Starston over Pulham North Green to Wacton Common (195 ft.), and in Suffolk, between Stradbrook and St. James', South Elmham, where the ground rises to 186 ft. From these points there are ever-varying gradations until the low level of the marshes, which bisect the district, is reached, where to the west the elevation is 67 ft., and to the east only 25 ft.

A mention of the height above the sea of some of the more prominent places will not be without interest to those who are familiar with the locality. As given by the Ordnance Survey of 1884, they are as follows, with the omission in a few cases of fractional parts: Harleston Market Place, 91 ft.; Redenhall Church, 89 ft.; Wortwell Green, 50 ft.; Alburgh Church,

140 ft.; Topcroft Church, 155 ft.; Gawdy Hall, 120 ft.; Thorpe Abbots Church, 137 ft.; Hoxne Church, 125 ft.; Wingfield High Elm, 146 ft.; Wingfield Castle, 163 ft.; Shotford Hill, 118 ft.; Shotford Bridge, 58 ft.; Weybread Church, 152 ft.; Weybread Windmill, about 170 ft.; Fressingfield Church, 130 ft.; Mettfield Church, 153 ft.; Mendham Church, 52 ft.; Homersfield Church, 48 ft.; St. Margaret's Church, South Elmham, 111 ft.

Through the low marshes before-mentioned the River Waveney, which divides the counties of Norfolk and Suffolk, traces its winding course from Lopham Ford, where it rises only 300 yards distant from the source of the Little Ouse, to the German Ocean at Yarmouth. In its present condition it is fed by all the streams or "becks," as they are locally called, of the neighbourhood, though, from the sudden fall of the valley-floor from Mendham to Earsham, a different arrangement of the watershed seems at one time to have existed.* The principal tributary is the stream which drains the shallow depression in which Harleston itself is situated. Others traverse the Denton, South Elmham, Weybread, and Hoxne valleys, and are known as the "becks" of the villages through which they respectively pass.

(i.) *Geology.*—A table of the geological formations, which appear at the surface in the district, is given on a subsequent page. The following short sketch will explain their local position and characteristics, the numbers having reference to the table, in which, as will be seen, the most recent beds are placed first. For many of the following particulars the writer is indebted to the Memoirs of the Geological Survey,† whose nomenclature is adopted; also to Dr. J. E. Taylor, of Ipswich, Mr. Charles Candler, and Dr. J. J. Raven for notes and suggestions.‡

The chalk, which covers the northern and western parts of the county and approaches the surface at Beccles, is nowhere visible, though it is reached by wells at 138 ft. (Hoxne), 55 ft. (Billingford), and possibly at 23 ft. (Wortwell). The exposed beds are confined entirely to the late Tertiary and Post-Tertiary formations.

No. 7. *Upper Crag.*—This is the oldest formation apparent

---

* Page 15.

† *The Geology of the Country around Harleston,* by Whitaker and Dalton, 1887; Eyre and Spottiswoode; 1s. The principal part of the district is included in the Geological Survey maps, 50 N.E. and 66 S.E.; Stanford; 3s. each.

‡ As the present sketch is intended for general readers, scientists are referred for complete information to the works of Messrs. Wood, Rome, and Harmer.

in the Harleston district, and is very limited in extent. It flanks the hill-side north of the Waveney from Homersfield Station up the valley of the Denton stream as far as the Beck Gate, and also appears on the escarpment known as the Bath Hills, Ditchingham. To the same formation may perhaps be referred the narrow strip of gravelly sand which forms the steep bank of Flixton Long Plantation, and beds of a tawny sand visible on the slope north of Harleston Bridge and occasionally in the floor of the Allotment Pit to the west. It abounds in fragments of shells, and is sometimes dug for paths.

No. 6. *Pebbly Beds.*—These include formations lying between the Crag and Glacial series, and possessing various local peculiarities. The chief section in the district is obtained in the Withersdale and adjacent valleys. A clayey bed, visible at the Cross Roads, is referred by Mr. H. B. Woodward [*] to the Chillesford formation, which is placed at the top of the Crag series; but a pit sunk a little time ago *below* this clay pierced a bed of sand full of smooth round pebbles of equal size, bearing a close resemblance to the Westleton Pebble Bed.

The Upper Crag and Pebbly formations are supposed to have been effected in a shallow sea at the mouth of a large river. As before said, they form but a small part of the features of the neighbourhood, which are almost entirely confined to the clays, gravels, and sands of the Glacial series.

The Glacial formations consist of beds of drift borne down from previously existing rocks by the action of ice. They are rendered particularly interesting by the proof which their composite nature affords of the altered climatic conditions of our temperate zone during the period of their deposition. Stones and boulders of a considerable size, oftentimes scratched and worn by long travel and the friction of the moving ice, are found resting in confused masses of sand and clay, with fossil remains of earlier ages. Some of these erratic blocks show, by their mineral composition, that they have been transported from the mountains of northern Europe, probably by the great Scandinavian ice-sheet, which filled the German Ocean at this time, and deposited its burden on the shores of Norfolk, as well as over parts of Denmark and Germany.

By far the larger portion, however, of the East Anglian drift has been derived by the action of land-ice from the rocks of Scotland and northern England, with a preponderance of chalk from the adjacent cretaceous formations. Among the rocks represented in this drift are the white sandstone and carboniferous limestone of the Pennine chain; the magnesian

---

[*] *Geology of England and Wales,* p. 467.

limestone of Durham and Yorkshire; the new red sandstone of Lancashire, Cheshire, and the Western Midlands; the oolitic limestones and clays of the northern and central plains; the pink granite, gneiss, and quartzite of the Highlands; the basalt of the heights of mid-Scotland, and perhaps of north-east Ireland; and the greenstone, greywacke, pink syenite, felstone, chert, porphyrite, Lydian stone, and porcellanite of the lower Silurian beds of south Scotland and Cumberland. These fragments seem to imply the existence of a large ice-stream passing from the region of the Grampians in a south-easterly direction over the Yorkshire wolds to the coasts of Norfolk and Suffolk, where, coming in contact with the more powerful ice-sheet of the German Ocean, it was compelled to diverge to the south-west until it melted in the valley of the Thames.*

The Glacial series is divided into three principal beds.

No. 5. *Lower Drift.*—This is also known as the Lower Boulder Clay. It is often difficult to separate from the succeeding Middle Drift, but it may be seen at the bottom in the Weybread Brickyard, and perhaps in a recently-opened pit at the foot of Needham Hill. Its contorted character was well shown formerly in a pit at Starston, near the Rectory.† Sections are exposed in pits near Denton Church; and on the Bath Hills the three beds of the series are distinctly traceable. The Lower Drift generally consists of a brown sandy loam, with erratic stones and but little chalk. It is particularly serviceable for brick-making.

No. 4. *Middle Drift.*—This is generally exposed on the sides of the valleys of the Waveney and its tributaries, where it is not overlaid by the river gravels. A good section of its fine sands is visible in the pit at Mendham on the Withersdale Road, and in the large pit on Shotford Heath, where they may be seen covered immediately by Post-Glacial gravel. The Middle Drift was probably deposited in very shallow water, subject to strong currents, and under a climate milder than that of the Lower and Upper Periods, which caused the glaciers to retreat for a time. It is almost entirely destitute of organic remains, and its chief economic value is the water supply of the district.‡

---

* Geikie, *The Great Ice Age*, ch. xxix.
† *Geological Magazine* (1868), p. 454.
‡ "The junction at Potter's Pits, Weybread, between this formation and a loamy clay below is most interesting. The water filters through the sand till it reaches the underlying and impervious bed, at which point it trickles out on the face of the section. A vigorous growth of coltsfoot (*Tussilago Farfara*), which is cut off in a clear line at the top of the clay where it meets the sand, shows the junction even where the slope is completely grassed."—C. C.

No. 3. *Upper Drift.*—The Upper Boulder Clay covers all the high lands to a considerable depth, and forms the principal feature of the neighbourhood. Its stiff plastic substance is employed in brick-making, and in the manufacture of "clay lumps," a local process in which it is mixed with chopped straw and dried in the sun, after the manner of the ancient Egyptians; as marl, it is used for manure on cultivated lands. Owing to atmospheric influences the clayey element in this formation is often washed out, leaving a thin layer of stony gravel and sand on the surface, which alters the apparent character of the district, as at Starston Brickyard. Fossils, derived from older formations and chiefly from the Chalk, Oolite and Lias, are common, and in some cases well preserved. Mr. Candler has obtained from the drift in this neighbourhood vertebræ of *Ichthyosaurus* and *Plesiosaurus*, and examples of the following genera of echinoderms and mollusca:—*Ananchytes, Micraster, Ammonites, Belemnites, Gryphæa, Inoceramus, Ostrea*, and others.* The clay contains manganese and a good deal of iron, which sometimes appears in the form of large nodules, and percolates through the underlying sands and gravels, giving them a reddish-brown colour.

The remaining two formations belong to the Recent and Post-Glacial Periods,† and are the earliest in which unquestionable traces of man's existence have at present been discovered.

No. 2. *Ancient Valley Gravel.*—This is the older, and contains bones of animals now extinct. It is found at various points in the vicinity of the Waveney valley, sometimes with traces of river loam or brick-earth, which is worked for brickmaking, as at Hoxne. It was probably formed, however, by the action of rivers before the existing physical features of the district were developed, and when the valleys were full of melting ice consequent on the change of temperature. In those early times, as Messrs. Whitaker and Dalton suggest, "the brooks from the South Elmhams, Metfield, and Fressingfield may have fed, not the Waveney, but the Little Ouse, as indicated by the westward trend of their channels, and by

---

* "Examples of the above-named genera occur in the Redenhall Road Brickyard pit. *Gryphæa incurva*, a shell from the Lias, is perhaps the most abundant and characteristic fossil of the Upper Boulder Clay at Harleston."—C. C.

† "I think that many of our so-called 'Post-Glacial' beds are *Post*-Glacial only in the sense of being more recent than the Chalky Boulder Clay. The latter formation, however, only marks the climax of the 'Great Ice Age,' and we know from the 'Purple Boulder Clay,' and 'Hessle Boulder Clay' of Yorkshire, that twice since that climax northern England has been buried beneath an advancing ice-sheet. To one or other of the two intervening temperate periods some at least of our 'Post-Glacial' brick-earths and gravels will perhaps be eventually assigned."—C. C.

the slight difference of level between Mendham and Lopham Ford." *

The late Mr. James Muskett, of Harleston, obtained from the gravel at Wortwell, west of Homersfield Station, a tooth of the mammoth (*Elephas primigenius*), the horn-core of an extinct ox (*Bos priscus*), beside bones of the elk and other animals. Beds of apparently the same age occur at Homersfield village (with animal remains), at Weybread Brickyard and Needham Hill (once perhaps continuous), and at Hoxne. As early as the year 1797, implements of chipped flints, evidently worked by man, were discovered at Hoxne, lying in undisturbed soil about ten feet below the surface, and *under* the bones of the elephant and ox. This fact, with similar discoveries in other parts of Europe, gives to this ancient gravel a peculiar anthropological interest.†

No. 1. *Recent River Deposit.*—As such must be classed the old gravels of Shotford Heath, Flixton, Bungay Common, Earsham, Brockdish, Thorpe Abbots, and Billingford, with isolated and later patches in the bed of the present valley, as at Mendham Old Priory and Wortwell. These gravels are of various ages, the highest levels as a rule being the oldest. They are due to the existing water-shed, having been deposited by river currents over the clays and sands of the Drift, now on one side of the valley, now on the other—rarely on both at the same time—in the form of indistinct terraces, as the river eroded its floor and changed the character of the district from that of a broad brackish estuary to the ordinary conditions of a narrow fresh-water stream.

It is well known that in historical times the Waveney was navigable above Harleston for small vessels, though now this is rendered possible only for barges by artificial means as far as Bungay, seven miles below. Anchors and traces of navigation have been found in the bed of the stream at Hoxne; and in the reign of King Stephen, when St. Mary's Priory was established at Mendham, the present site of the ruins in the marshes was called by a distinct name, Hurst or Bruninghurst —probably from the coppice of alders which then stood on it —and is described in the founder's deed as an *island* ; ‡ while, according to Suckling, as late as the year 1549, during Kett's rebellion, a small pinnace was prepared at Yarmouth to carry twenty men up the river as far as Weybread.§

---

* *Memoir of the Geological Survey of Harleston*, p. 27.
† *Cf.* Lyell, *Antiquity of Man*, p. 217 ; Lubbock, *Prehistoric Times*, p. 359.
‡ Dugdale, *Monasticon*, vol. v., p. 56. In the *Gentleman's Magazine* (1808), p. 969, a plate shows the chapel and prior's lodge then standing, though partially in ruins.
§ Introduction to the *History of Suffolk*, p. 7. Suckling's authority is Swinden's *History of Great Yarmouth* (1772), but I fear he has arrived at too

Even the present marshes, however, are gradually losing their character by drainage. They consist of vegetable mould, gravel, sand, and mud borne down by the river and its streams from the adjacent uplands, and are, at their greatest width, three-quarters of a mile across. They mostly produce hay-grass, and occasionally afford a rough grazing-ground for cattle.

GEOLOGICAL FORMATIONS OF THE DISTRICT.

FORMATIONS.	CHARACTER.	PERIOD.	CLIMATE.
ALLUVIAL.			
1. River Deposit.	Mud, gravel, sand, and débris.	Recent.	Temperate.
2. Ancient Valley Gravel.	Gravel with flint implements, bones of extinct animals, and occasional brick-earth.	Pleistocene.	Unsettled.
GLACIAL.			
3. Upper Drift.	Stiff chalky clay with erratic boulders and derived fossils.		Arctic.
4. Middle Drift.	Fine sand and gravel, with false bedding.		Mild.
5. Lower Drift.	Brown sandy loam, contorted, rarely with chalk fragments.		Sub-Arctic.
LITTORAL.			
6. Pebbly Beds.	Sand with round pebbles; sometimes a loamy clay.	Pliocene.	Cold.
7. Upper Crag.	Shelly tawny sand, charged with iron.		Temperate (tending in the lower formations to Tropical).

(ii.) *The Stone Age.*—Allusion having been made in the Geological sketch to the ancient flint implements found at Hoxne, a short account of some of the forms met with in other parts of the district will not be out of place. The following notes have kindly been made for the purpose by Mr. John A. Holmes from his own and his brother's observations:—

"The implements and weapons of stone picked up in the

---

hasty a conclusion. The passage in Swinden (p. 939) occurs in a transcript of the orders sent to the Yarmouth garrison for the defence of the town against the rebels. Amongst the precautions is this:—"Item, that a small pinnace do go up to Waybridge, being victualled for four days, having twenty-six men in her." This, I presume, is not Weybread near Harleston, which was supplying men and money for Kett and his followers, but Weybridge at Acle on the Bure, about nine miles from Yarmouth.

B

neighbourhood of Harleston are almost exclusively of the older or rough-chipped type, the ground and polished implements of the Neolithic Period being extremely rare. Of the following specimens four are in my own collection; it will be observed that there is no record of a Neolithic *weapon* :—

POLISHED IMPLEMENTS.

TYPE.	CONDITION.	MATERIAL.	LOCALITY.
Axe	Perfect	Dark flint	St. Margaret's.
Axe	Butt end	Dark flint	St. Margaret's.
Axe	Butt end	Dark flint	St. Margaret's.
Axe	Butt end	Chert	Flixton.
Hammer*	Perfect	Quartzite	Harleston.

"The rough-chipped implements and weapons discovered in the district consist of axes of various descriptions, scrapers, spear, javelin, and arrow-heads, all of flint, many very roughly worked, and stained or encrusted with the material in which they have been resting since their disuse. Whether all these older implements should be referred to the extreme antiquity of the Hoxne flints it is difficult to say, as they have been picked up in most cases on the surface of the land, on stone-heaps, or in the bed of streams.† My brother, however, discovered in a newly-opened pit at Brockdish an implement *in situ*, which may probably have been wielded by Palæolithic man in times when the mammoth and woolly rhinoceros, whose bones lie buried in the ancient gravels of our valley, shared with him the struggle for existence.‡ This specimen is singularly worked, and evidently intended for fighting purposes. It is nearly semicircular, the entire edge eight inches in length, and with a natural hole through the substance of the stone towards the back, which could be utilised for passing a thong through in hafting.

"Space will not permit a complete list of all the rough-chipped flints found in the neighbourhood. The following are selected types of these older implements in my own possession, most of them nearly perfect :—

ROUGH-CHIPPED INSTRUMENTS.

TYPE.	MATERIAL.	SIZE.	LOCALITY.
Battle-axe	Grey flint	6 in. × 4 in.	Pit, Brockdish.
Battle-axe	Dark flint	5 × 2½	Stoneheap, St. Margaret's.
Adze	Dark flint	4 × 2½	Roadside, St. M.
Spear-head	Grey flint	4½ × 2	Ploughed field, St. M.

* In the possession of the Rev. C. R. Manning, Rector of Diss.
† Evans, *Ancient Stone Implements of Great Britain*, pp. 60—77.
‡ *Cf.* Wilson, *Prehistoric Man*, pp. 26—28.

ROUGH-CHIPPED INSTRUMENTS (continued).

TYPE.	MATERIAL.	SIZE.	LOCALITY.
Spear-head	Grey flint	3 in. × 2 in.	Ploughed field, St. M.
Spear-head	Dark flint	4½ × 2	Brook, St. M.
Arrow-head	Grey flint	2 × 1	Ploughed field, St. M.
Arrow-head	Grey flint	1½ × ¾	Ploughed field, St. M.
Scraper	Grey flint	Nearly circular, 2½	Ploughed field, St. M.
Scraper	Grey flint	Nearly circular, 1½	Ploughed field, St. M.

JOHN A. HOLMES, M.A."

(iii.) *Climate.*—As the climate of a district naturally affects the plant life, and is, in its turn, dependent on the physical features, an account of the Flora of Harleston will not be complete without a brief summary of its meteorology. Full particulars are obtainable through the careful and long continued observations made by the Rev. Charles Gape, M.A., (late) F. Met. Soc., at Rushall Vicarage, who has kindly placed his records in the writer's hands to use as most desirable.

From these observations we find that the average temperature of the Harleston neighbourhood is a little below that of Norwich, probably owing to the open nature of the country. The following analysis of the temperature for three years shows the lowest and highest reading of the thermometer (in the shade), and the mean temperature of day and night combined. The instruments are exposed.

TEMPERATURE.

	1884.			1885.			1886.		
	Lowest.	Mean.	Highest.	Lowest.	Mean.	Highest.	Lowest.	Mean.	Highest.
Jan.	26°	41°	54°	17°	34°	53°	19°	35°	52°
Feb.	19	40	54	17	41	57	17	33	45
Mar.	20	43	69	18	37	58	20	38	62
Apr.	20	44	67	18	45	74	25	46	70
May	29	54	78	22	49	76	24*	52*	77*
June	36	55	81	30	57	83	33	56	81
July	37	62	87	35	60	87	34	62	87
Aug.	36	63	90	31	57	79	40	62	85
Sep.	31	58	81	25	54	76	36	59	88
Oct.	24	50	65	29*	45*	58*	38*	55*	76*
Nov.	21	40	59	26	41	58	27*	43*	55*
Dec.	19	38	55	14	37	51	11	34	54

The average heat on a summer day at Harleston is about 75°; on a winter day it is about 42°. The average cold on a

* Record imperfect.

summer night is 47°; on a winter night, 31°. The highest temperature recorded in the sun during the years 1884—86 was 107° on Sept. 18th, 1884, though greater heat was experienced on August 11th of the same year, when the thermometer recorded 90° in the shade.

Thunderstorms are not infrequent, though often confined to certain water-sheds; the tremor of the earthquake which took place at 9·20 a.m., on April 22nd, 1884, at Colchester, was distinctly perceptible.

A comparison of the mean summer (July) and winter (January) and annual temperatures (day and night combined) with those of well-known places in Great Britain, and also of Continental cities having the same latitude as Harleston, will show the relative nature of its climate:—

	SUMMER.	WINTER.	ANNUAL.
Harleston	62°	38°	49°
London	63	37	51
Land's End	64	43	54
Edinburgh	59	37	47
Amsterdam	63	33	47
Berlin	66	27	48
Warsaw	64	22	46

From this calculation it will be seen that though the mean annual temperature of Harleston is almost the same as that of Berlin, yet the variation of the seasons is less marked; there is only 24° difference between summer and winter at Harleston, as compared with 39° at Berlin. The cause is its insular and quasi-littoral position; but if we compare the climate with that of an *English* inland town of the same latitude, we find that, owing to the stiff soil, the absence of forests and hills, and the prevalence of the east and north-east winds during the early months of the year, the mean annual temperature of Harleston is not above the British average for the same latitude, notwithstanding its proximity to the moderating influences of the sea.

Rain, including snow, hail, and heavy mist, usually falls in the district on a little under half the number of days in the year, and to the comparatively small amount of 26·23 inches, according to the highest average. The daily records are consequently low, and the fall of 2·21 inches on Sept. 4th, 1884, was almost unprecedented, as very rarely an inch is attained in one day. The annual fall in London is estimated, on the highest average, at 27 inches; in the higher tracts of Wales it is over 100 inches; in the Cumberland Lake District over 140 inches; and in the hills of north-eastern India the yearly average is 600 inches, of which 500 inches falls in the seven months' monsoon.

Although the rainfall at Harleston is among the smallest in England, the district is liable to frequent and extensive floods, owing to the slight gradient of the Waveney valley and the numerous obstacles which impede the course of the river. Snow generally covers the country in the early part of the year, and often in the later months. In the spring the intensely keen north-east wind, blowing from the frozen shores of Scandinavia, lowers the temperature and retards vegetation, the flowering of plants being about three weeks later than in the south of England.*

The subjoined tables show the monthly rainfall for the last five years, with monthly and yearly averages for ten years, and the number of days in each month on which rain (0·01 inch) usually falls on both sides of the Waveney valley. An average for the last five years is also given as perhaps the truer estimate, because the seasons have been more normal, especially in contrast to the extraordinary humidity of previous summers. The record for June of the present year (1887) is, however, unusually small, after the first two or three days the month being practically rainless; and the year, as a whole, has been particularly dry.

Of the two records here tabulated, the northern was kept by the Rev. C. Gape, of Rushall Vicarage, at 117 feet above sea-level; the southern by the Rev. J. H. White, of Weybread Vicarage, at 152 feet above sea-level. The lower average fall in the southern station is probably due to the fact that it is sheltered by higher ground from the beat of the rain-bearing

---

* The following conditions of climate during the years immediately preceding 1883 are worthy of record:—

January, 1879.—A severe frost, followed by a rapid thaw, causing an extensive and long-continued flood in the Waveney valley.

August 2, 1879.—A thunderstorm, with immense hailstones, destroying trees, glass, and roads. Rainfall at Rushall, 2·48 inches!

November and December, 1879.—Extreme cold, the thermometer on the ground registering 26° of frost on several occasions.

January 18, 1881.—A great gale, and heavy fall of snow to the depth of one foot in the streets, followed by a severe frost. All roads blocked, and a train embedded in a drift at Pulham.

July 15, 1881.—Extreme heat, the thermometer registering 95° in the shade.

October 10, 1881.—A great gale, with heavy rain, destroying trees, &c.

October, 1882.—An unprecedented rainfall of 6? inches for the month. The annual amount was 34·55 inches (Rushall); much above the average.

winds. Both records are published annually in Symons' *British Rainfall*.

RAINFALL AT RUSHALL, NORFOLK.

| | 1883. | 1884. | 1885. | 1886. | 1887. | Average Fall. | | Av. No. of Days. |
						5 yrs.	10 yrs.	
	in.	in.	in.	in.	in.	in.	in.	
Jan.	1·54	1·52	2·26	2·11	1·33	1·75	1·41	15
Feb.	2·16	0·66	2·22	0·26	0·59	1·18	1·62	13
Mar.	1·55	1·10	1·06	1·21	1·58	1·30	1·27	13
Apr.	0·83	1·50	1·31	1·12	1·07	1·16	1·50	13
May	1·63	1·08	2·66	1·76	2·02	1·83	1·93	13
June	2·70	0·90	0·73	0·49	0·22	1·00	1·65	11
July	3·34	1·65	0·96	3·52	1·38	2·17	2·90	15
Aug.	0·71	1·36	1·07	1·68	1·45	1·26	2·55	12
Sep.	3·18	4·19	5·14	1·69	1·88	3·21	2·97	17
Oct.	3·20	3·13	5·45	2·03	2·75	3·31	3·38	18
Nov.	3·52	1·79	3·09	2·45	2·29	2·63	2·80	18
Dec.	2·15	2·50	0·87	3·51	1·41	2·09	2·25	17
Total	26·51	21·38	26·82	21·83	17·97	22·89	26·23	175

RAINFALL AT WEYBREAD, SUFFOLK.

| | 1883. | 1884. | 1885. | 1886. | 1887. | Average Fall. | | Av. No. of Days. |
						5 yrs.	10 yrs.	
	in.	in.	in.	in.	in.	in.	in.	
Jan.	1·55	1·35	1·88	2·07	1·29	1·63	1·45	14
Feb.	1·97	0·52	2·23	0·18	0·49	1·08	1·48	12
Mar.	1·08	1·07	0·95	1·26	1·34	1·14	1·15	11
Apr.	0·80	1·26	1·18	1·27	1·37	1·18	1·41	11
May	1·50	0·89	2·76	1·77	1·89	1·76	1·71	11
June	2·24	1·12	0·95	0·48	0·08	0·97	1·62	9
July	3·20	1·72	1·17	3·29	0·96	2·27	2·57	13
Aug.	0·46	0·92	0·94	1·29	1·59	1·04	2·16	10
Sep.	2·21	3·57	5·34	1·05	2·03	2·86	2·84	14
Oct.	2·83	2·76	5·37	1·80	2·20	2·99	3·07	15
Nov.	3·12	1·53	2·80	2·76	2·15	2·47	2·68	16
Dec.	1·76	2·46	0·96	3·65	1·00	1·97	2·13	15
Total	22·72	19·17	26·53	20·87	16·39	21·36	24·27	151

B.—BOTANICAL RESEARCH.

Though Harleston itself can claim but few native botanists, yet the district has not remained unnoticed or unworked. Little, it is true, has hitherto been recorded for its northern and western limits, but on the east and south observations have been carried on for a period of over eighty years.

(i.) *Past Observers.*—Attention was first called to the floral characteristics by Mr. T. J. Woodward, F.L.S., who for a long time resided at Bungay, and supplied information to the well-known *Botanist's Guide* of 1805, and to the later editions of Withering's *Arrangement of the British Flora.* Additional observations were furnished by him to the *New Botanist's Guide* (1835), together with those of a younger botanist, resident in the same town, Mr. Daniel Stock. Mr. Stock's records are deprived of much of their value for our present purpose by their vagueness, as in most cases his various localities are included under the name of the town in which he lived. Such as they are, however, they supply the chief information of the neighbourhood in Henslow and Skepper's *Flora of Suffolk,* published in 1860. Mr. Stock furnished additional notes to the Rev. Kirby Trimmer, who in 1866 edited the result of his own inquiries and observations in the county of Norfolk, and has brought them up to date by a Supplement published a short time ago.

Meanwhile, a careful examination of the country around Harleston was being made by the Rev. E. A. Holmes, F.L.S., late Rector of St. Margaret's, South Elmham. Commencing his observations on his institution to the living in the year 1833, he continued them for more than fifty years, keeping an annual record for at least half that period. Unfortunately, in this case also, an absence of specified localities detracts in a great measure from the extreme value of his work. This deficiency has been somewhat counteracted by the personal knowledge which it was the writer's privilege to have of Mr. Holmes. During many botanical rambles and conversations opportunity was given for identifying the localities of the rarer plants, and even of verifying some of the records of Mr. Stock. The notes thus made, with a few written *suâ manu* in an interleaved copy of the *Botanist's Guide,* have proved of great assistance in determining the localities of other plants in the annual lists, which, through the kindness of Mrs. Holmes, have been inspected for the purposes of this Flora. Mr. Holmes' long residence in the neighbourhood, his complete knowledge of the subject, and intimate acquaintance with the immediate vicinity of Harleston, render his records by far the most valuable of the past.

Two names connected with the town itself deserve a mention—that of the Rev. H. Tilney, who occasionally resided here in the early part of the present century, and contributed many localities of rare plants to the *Botanist's Guide;* and that of the late Mr. James Muskett, whose entomological pursuits brought him into close contact with the wild flowers, and enabled him to give much interesting and trustworthy information.

(ii.) *The Harleston Botanical Club.*—In the autumn of the year 1882 the writer became acquainted for the first time with this Eastern county. Having made some personal observations of the Flora during the following year, on his suggestion it was resolved, in the spring of 1884, to form a small club of working botanists resident in the town and neighbourhood for the purpose of collecting information upon its flowering plants and ferns. A short account of methods used and results attained may perhaps suggest to other lovers of nature the formation of a similar parochial society for the recreation and instruction of its members, as well as for the general advancement of science. Nor was there anything in the existing local circumstances which predicted or insured the success of the Harleston Club; on the contrary, from the first the difficulty of combined work was foreseen, owing to the various occupations of the observers. Interleaved copies of the *London Catalogue of British Plants* were, however, supplied for independent use, while the members offered to share in common such books and knowledge of the subject as they possessed for the identification of doubtful discoveries. At the end of the year the catalogues were collected, and their contents tabulated in one schedule, which was circulated amongst the workers. Such is the course which has been pursued in the main for the four years during which the Club has existed, though, as the Flora has been more completely recorded, fewer annual additions have required shorter schedules. Apart from frequent walks in company, two general meetings have usually been held each year, at which results have been announced, reports collected, and plans for extended observations submitted and accepted. In order to encourage a systematic study of botany, the Club, for two successive years, offered prizes at the Harleston Horticultural Society's shows for specimens of plants belonging to the natural orders *Rosaceæ, Compositæ, Juncaceæ, Cyperaceæ, Gramineæ,* and for aquatic plants, but the response did not warrant their continuance.

In the first year (1884) the area under observation was

that included by a circle of four miles' radius from Harleston Station. At the close of the year 532 species of flowering plants and ferns were recorded, with twenty varieties. For the next year (1885) the circle of observation was extended to a radius of five miles. Eighty additional species were recorded, including a more complete study of the local *Rubi* and *Gramina*. In the following year (1886) the area was extended another mile, and forty new species were added, with five new varieties. During the present year (1887) a general revision of the existing area has been attempted, and, as might have been expected from the active investigations of previous years, few additions have been made, twenty-four new species only having been recorded.

In order to render the details of the Flora more perfect and representative, the observations of Mr. Holmes and others before-mentioned, together with the reports of friends, *duly examined and credited*, have been included in the present account. The following table will therefore show the number of species at present known within six miles of Harleston, and in those parishes through which the six-mile circle passes. Eighteen of their number are probably extinct, not having been observed for many years:—

FLOWERING PLANTS, FERNS AND ALLIES.

Species observed by the Harleston Botanical Club (1884)	532
,, ,, ,, ,, ,, ,, (1885)	80
,, ,, ,, ,, ,, ,, (1886)	40
,, ,, ,, ,, ,, ,, (1887)	24
Total Species observed by the Club	676
Species added by an examination of Printed Records ...	50
,, ,, ,, ,, Manuscript Lists ...	57
,, ,, ,, ,, Credited Reports ...	8
Total Species within a six-mile radius	791
Additional Species already recorded within an eight-mile radius	43
Total	834

(The species are determined by the *London Catalogue*, 8th edition, 1886.)

A comparison of the Harleston Flora with those of Great Britain and the Eastern Counties will be found in the next section of the Introduction. The proceedings of the Club have already been noticed in a paper read before the Norfolk and Norwich Naturalists' Society in February, 1886, and followed by an additional paper in the present year.* The

---

* *Transactions of the Norfolk and Norwich Naturalists' Society*, Vol. IV., Part II., p. 225, and Part III., p. 395.

enumeration given above differs slightly from that recorded in the Transactions of the Society, and supersedes it.

The progressive aspect of the Club's efforts will be shown by a mention of the species which it has been enabled to add to the Floras of Norfolk and Suffolk. The list of Norfolk plants, with additions published yearly, is already approximately complete, under the direction of a friend interested in the Club, Mr. H. D. Geldart of Norwich. The Suffolk Flora is very imperfect, so far as the present published records are concerned. Even Britten's list of Suffolk plants (1874) admits a great many additions from the Harleston list, but it has been considered hardly just to base calculations on accounts so defective. The Rev. W. M. Hind, LL.D., of Honington, near Bury, has prepared for immediate publication a new Flora of the county. The manuscript of the present list has therefore been submitted to him, and, in addition to several records of plants collected by Dr. Hind himself, the writer has received a statement of those of which the Club furnishes the first information. By this means it is hoped no undue credit is gained.

SPECIES AND VARIETIES RECORDED FOR THE FIRST TIME.

IN NORFOLK.
SILENE NUTANS.
VIOLA REICHENBACHIANA.
RUBUS SALTERI.
RUBUS SCABER.
ORNITHOGALUM PYRENAICUM.
CAREX ACUTA (gracilesceus).
*ANEMONE APENNINA.
*ERANTHIS HYEMALIS.
*TRIFOLIUM HYBRIDUM.
*PETASITES FRAGRANS.
*NARCISSUS BIFLORUS.
*LILIUM MARTAGON.

IN SUFFOLK.
RUBUS RHAMNIFOLIUS.
RUBUS KOEHLERI (infestus).
JUNCUS DIFFUSUS.
SPARGANIUM NEGLECTUM.
*LILIUM MARTAGON.

C.—FLORAL CHARACTERISTICS.

Under this title an attempt has been made to show, by comparison with other Floras, some of the peculiarities of the botany of Harleston. In so small an area it is hardly possible that great divergencies can exist, but the eastern position of the district, its situation on the sands and clays of the Drift, and its proximity to the sea-coast, give to it a character interesting if not unique.

(i.) *The British Flora Compared.*—The late Mr. H. C.

* Introductions.

Watson, in his *Cybele Britannica*, was one of the first to systematise the distribution of plants in the British Islands. Considering the Flora, first of all, with reference to the climate, he arranged it under two regions, called respectively the Agrarian and the Arctic, each containing three zones, rising in altitude, and distinguished by the presence or absence of certain well-known plant forms. Following his arrangement, the lowest Agrarian zone is marked by the presence of the southern-type plants, *Clematis vitalba*, *Rubia peregrina* and *Cyperus longus*; the mid-Agrarian zone by the absence of these species, but the presence still of *Rhamnus catharticus*; the highest Agrarian zone by the absence of *Rhamnus*, but the presence of *Pteris aquilina*. The lowest Arctic zone is, in its turn, distinguished by *Erica tetralix*, without *Pteris*; the mid-Arctic zone by *Calluna vulgaris*, without *Erica*; and the highest Arctic zone by *Salix herbacea*, without *Calluna*.*

As will be seen from the first record in the Flora, the district of Harleston lies in the lowest Agrarian zone; though *Rubia* and *Cyperus* are not found in the neighbourhood, *Clematis* is decidedly frequent. In fact, all the country south of the Humber and the Dee, or an imaginary line drawn from Liverpool to Hull, is included in this lowest zone, except the mountainous tracts of Wales, and the high moors of the Severn provinces. The following plants found in the Harleston district are, in most cases, generally distributed throughout the zone to which it belongs, but are unknown as *natives* in the mid-Agrarian zone immediately above it: *Helleborus fœtidus, Aconitum napellus, Trifolium glomeratum, Trifolium suffocatum, Lathyrus aphaca, Tillæa muscosa, Œnanthe fluviatilis, Fœniculum vulgare, Galium anglicum, Linaria spuria, Chenopodium hybridum, Rumex pulcher, Carpinus Betulus, Ruscus aculeatus, Fritillaria meleagris, Ornithogalum pyrenaicum, Tulipa sylvestris, Alopecurus fulvus;* while the following *denizens* are absent in the succeeding zone: *Adonis autumnalis, Erysimum cheiranthoides, Verbascum Blattaria, Setaria viridis.* No mention is here made of the more local species which are enumerated under the next section.

Mr. Cottrell Watson then proceeds to resolve the British Flora into types with reference to geographical position.

1. *The British Type* includes thirty-four species of plants thoroughly native to our island, though not to be considered necessarily of sole British origin. All of these the Harleston Flora possesses.

2. *The English Type* includes thirteen species adapted to

---

* *Cf. Cybele Britannica*, vol. i., p. 40; *Compendium of the Cybele*, pp. 14—32.

the geographical position of England.* Of these the district claims all except *Ulex nanus*, which, however, is reported as growing a few miles beyond the border.

3. *The Scottish Type;* and 4. *The Highland Type* embrace species generally unknown in lower latitudes. It is noticeable, however, that the district possesses two species of Ferns which are natives of higher elevations—*Asplenium viride* and *Cystopteris fragilis*. There is, of course, some difficulty in deciding how far they owe their presence here to human agency, but both have been recorded by various observers for nearly half a century, and in 1884 *Cystopteris* appeared—and was immediately eradicated in spite of the efforts of the Club to preserve it—in a new locality in which its intentional introduction was out of the question.

5. *The Germanic Type* embraces eleven species of plants having a tendency to a distribution connected with the provinces of England bounded by the German Ocean and North Sea.† It is natural that, owing to our eastern position, a record of all these species should be expected; but, owing also to the absence of the chalk which covers a large part of the east of England, and the distance which now separates the district from the sea, only four species are forthcoming: *Reseda lutea, Silene noctiflora, Lactuca scariola,* and *Aceras anthropophora.*

6. *The Atlantic Type* includes species having a tendency to a distribution on the western side of the island. It is not to be expected that such species should be met with here, though *Ceterach officinarum*, which is generally known as a western plant, finds a place in the district list.

Hence it will be seen that the Flora of Harleston belongs to the lowest zone of the Agrarian Region; that it confirms the British, English, and Germanic character its geographical position assigns to it, and at the same time anticipates by certain marked forms the approach of higher latitudes.

(ii.) *The Flora of Norfolk and Suffolk Compared.*—Allusion has already been made to the new species which the Club has been enabled to add to the Flora of these counties; it, therefore, only remains to point out general affinities and divergencies between their botany and that of the district.

An inspection of our list reveals a marked absence of heath

---

* English Type.—Examples:—*Rhamnus catharticus, Ulex nanus, Tamus communis, Bryonia dioica, Hottonia palustris, Chlora perfoliata, Sison amomum, Linaria elatine, Ranunculus parviflorus, Lamium galeobdolon, Hordeum pratense, Alopecurus agrestis, Ceterach officinarum.*

† Germanic Type.—Examples:—*Frankenia laevis, Anemone pulsatilla, Reseda lutea, Silene noctiflora, Silene conica, Pimpinella major, Pulicaria vulgaris, Lactuca scariola, Atriplex pedunculata, Aceras anthropophora, Spartina stricta.*

and marsh plants from the district. The sole representative of the natural order *Ericaceæ* is one plant of *Calluna erica* (*vulgaris*). The so-called "heaths" of the neighbourhood are usually beds of Post-Glacial gravel, with disused pits, possessing none of the characteristics of true heather-land, except the gaunt forms of *Pinus sylvestris*. It is, moreover, strange that species so generally distributed as *Saxifraga tridactylites*, *Pedicularis sylvatica*, *Polystichum angulare*, *Athyrium Filix-fæmina*, and *Asplenium adiantum-nigrum* should be amongst the local rarities; and when we turn to the Flora of the marshes, where the frequent occurrence of typical plants might be reasonably expected, we find that though records are given for some of them, yet, on the whole, they are now seldom seen. This is probably due to the effective system of drainage, which was established about thirty years ago; for in the "good old times" the Bladderworts, the Water Soldier, and many other interesting plants gladdened the eyes of the fortunate observer; but, while they still linger in higher parts of the Waveney valley and are to be found in the lower reaches of the river, they are known to us no more.

The district nevertheless is abundant in species belonging to the natural orders *Geraniaceæ*, *Leguminiferæ*, *Rubiaceæ* and *Scrophularineæ*. It is especially rich in the Monocotyledonous orders *Orchidaceæ*, *Irideæ*, *Amaryllideæ* and *Liliaceæ*, the chalky clay seeming to favour the growth of bulbous plants. Forty-six species belonging to these last-named orders are reported in the latest lists of Norfolk plants: thirty-seven species have already been observed in the neighbourhood of Harleston, including the two new plants *Ornithogalum pyrenaicum* and *Narcissus biflorus*, but exclusive of *Narcissus major* and *Asparagus officinalis*, established in a wild locality for nearly a century.*

As the surface soil of the district mainly consists of the chalky boulder clay before mentioned, we might expect to find that some of those plants which have a preference for the

---

* The habitats of these two plants—the large hedgebank on Beacon Hill above Shotford Bridge, and the clump of trees a short distance eastward, called Mendham Grove, Norfolk—are interesting. By their Flora they suggest the existence in former days of gardens, and tradition asserts that in the last century two halls, one of them perhaps never completed, stood on these spots. An old map of the year 1795, however, shows no such dwellings there, and the oldest inhabitant of Harleston, Mr. Barber, who remembers the locality as it was in 1810, can give no information respecting them. It is, perhaps, possible that at one or other place stood WHICHENDON, or WHITE-HILLS HALL, the family seat of the Frestons, to whom the manor was granted in the first year of the reign of Edward VI. The family, whose history is traced by Blomefield (*Hist. Norfolk*, Vol. V., p. 377), held an important position during the seventeenth and eighteenth centuries, but now not even the name—much less the site—of their ancestral mansion is remembered in the neighbourhood.

chalk formation occur also on its drift. Such have been recorded in *Reseda lutea, Galium tricorne, Carduus nutans, Cnicus acaulis, Lactuca muralis, Specularia hybrida, Aceras anthropophora, Ophrys apifera, Ophrys muscifera, Iris fœtidissima,* and probably in *Tulipa sylvestris, Bupleurum rotundifolium* and *Viburnum Lantana.**

Of plants chiefly confined to the counties of Norfolk and Suffolk the district possesses *Holosteum umbellatum, Trifolium ochroleucum, Veronica verna* (extinct?), *Primula elatior, Muscari racemosum, Potamogeton trichoides, Apera interrupta* and *Corynephorus canescens.*

The *London Catalogue* (8th edition) records 1,858 species of flowering plants, ferns, and allies, for the British Isles; the Flora of Norfolk, published by the Norwich Society, reports about 1,200 species; the new Flora of Suffolk will include 1,219 species, and 177 varieties; the Flora of Harleston records for six miles 791 species, and 36 varieties, and for eight miles, according to present information, 835 species.

(iii.) *Traces of a Maritime Flora.*—In an age of theories it is not the writer's wish to multiply them needlessly. There are, however, certain species of flowering plants growing in the higher Waveney valley which seem to exist as relics of an older and maritime Flora. In the Geological sketch† mention has been made of the former condition of the valley as a brackish estuary, and of the traces still lingering in its physical features. An observer standing on Redenhall Church tower during a period of flood will gain some idea of the ancient character of the surrounding country. In association with this estuarine condition, a class of plants incidental to salt marshes and the sea-coast was naturally established, and some of them seem still to linger. Such, for instance, are the following Umbelliferous species:—

Fœniculum vulgare.—According to Hooker's Flora (3rd edition), the Fennel is found on "sea-cliffs," and is "perhaps native from North Wales and Norfolk to Cornwall and Kent." In a corresponding position inland, upon the sides of the valley of the Waveney and its tributaries, this plant is frequent throughout the district. As early as 1835 it was recorded as growing "on the Bath Hills for many years," and its prevalence seems only satisfied by the supposition of a native origin, and not by an introduction from innumerable gardens.

Apium graveolens.—"Marshy places, chiefly by the sea" (Hooker). Though the Celery is not recorded at present within

---

* In this and the statements of the next section Hooker's *Student's Flora* (3rd edition) and Watson's *Topographical Botany* (2nd edition) have been adopted as standards.

† Page 16.

our six-mile circle, it grows just outside it at Bungay, and in the lower parts of the valley is frequent.

Smyrnium olusatrum.—"Waste places, especially near the sea" (Hooker). This plant is a doubtful native in Britain, and if only a few specimens of it occurred in the district, it would not call for notice, as in olden days it was cultivated as a pot-herb under the name of "Alexanders." It abounds, however, in many spots on the valley-sides, as a rule at a somewhat lower level than the Fennel.

The following plants are particularly interesting in this connection, as they are representative of the Flora now existing on the coast :—

Trifolium suffocatum.—This Trefoil grows in a pit on Bungay Common, and also at Broome. It is found "especially near the sea," and, though a rare plant, is abundant on the sandy Denes of Yarmouth, Lowestoft, and Southwold.

Erodium cicutarium.—The Stork's-bill, which is "most frequent by the sea," is common on dry banks throughout the district. Both here and on the coast it is one of the earliest of the wayside flowers.

Teesdalia nudicaulis.—The Teesdalia forms one of the principal elements of the coast Flora, and in the early months of the year characterises it. In the Harleston district it grows on a bank of Post-Glacial gravel, called "Homersfield Heath," opposite the Flixton Park gates, and has also been found on the gravel at Needham Hill and Ditchingham.

Senecio viscosus.—This Groundsel is especially noticeable on the coast, where its sticky stem is coated with the blown sand. In our district it has been observed in a gravel-pit at Ditchingham.

Corynephorus canescens.—This rare grass was discovered by Mr. Walter Cordwell, of our Club, on the Post-Glacial gravel at Flixton before-mentioned. It is plentiful on the Denes at Yarmouth and Lowestoft, but grows nowhere else in England. Its presence so far inland (about sixteen miles due west) is remarkable ; but Dr. Hind has received the report of another inland station in north-west Suffolk. The fact that, sixty years ago, Mr. Stock cultivated specimens of this plant in his garden at Bungay, does not seem sufficient to account for its establishment on Homersfield Heath. The east winds, which might perhaps transport the seeds from the coast itself, are almost entirely confined to the earlier months of the year, when the grass, which is an annual, is not even in bloom. The late Mr. Holmes, moreover, who knew Mr. Stock, and often spoke of Homersfield Heath, never mentioned its existence or its introduction there. As it grows in immediate association with *Teesdalia*, and in close proximity to the plants named above, it points its origin rather to the littoral conditions which once existed in its present locality.

Other plants there are which, by a like association, imply a similar condition; amongst them we may mention *Samolus Valerandi, Iris fœtidissima* and *Rumex maritimus.** Sufficient traces are, therefore, extant to bear common testimony with the physical features to the great change which has been taking place in the general aspect of the valley during the formation of its present Flora.

---

* Additional interest has been added to these conclusions by the capture of the rare moth *Eremobia ochroleuca*—at Harleston by Mr. C. Candler, and at Denton by the Rev. C. T. Cruttwell. Its few known haunts are on or near the sea-coast.

II.—BOTANICAL BOOKS AND COLLECTIONS.

Many who are desirous of acquainting themselves with our native flowers are occasionally in doubt as to the best books for their purpose. It has therefore occurred to the writer that a short and informal excursus on Botanical books, with a few hints to intending collectors, would be both welcome and useful.

There are, of course, a great number of treatises of recent publication bearing on the subject, and on special branches of it, but our intention is not to advise specialists or advanced students; the works enumerated are for general knowledge, and their value has been tested by practical experience. In order to include beginners, who might be deterred from taking up the pursuit through fear of long words and hard names, the list is arranged under two heads, according as the subject is treated in the (so-called) "popular" and "scientific" methods.

POPULAR TREATISES.

1. *The Flowers of the Field*, by the Rev. C. A. Johns, F.L.S.; post 8vo. 5s. S.P.C.K.

This is an excellent book, and its low price brings it within the reach of all. The letterpress, which gives an explanation of the structure of plants, contains a short account of most British species as far as the Pond-weeds and Rushes. It is copiously illustrated with wood-cuts, and has done more to diffuse a knowledge of plant-life than any other book of its kind.

2. *British Ferns and Allied Plants*, by Thomas Moore, F.L.S.; coloured plates; fcap. 8vo. 1s. and 3s. 6d. Routledge.

This little book is a useful adjunct to Johns' *Flowers of the Field*. Besides the plates, there are numerous illustrations of varieties, with full particulars of Fern structure and culture.

2*. *A History of British Ferns*, by Edward Newman, F.L.S.; 8vo. 18s. Sonnenschein. An abridged edition, 2s.

The excellent engravings which distinguish Newman's works surpass the coloured plates of most authors. The illustrations of varieties are very numerous and minute in their details.

3. *Familiar Wild Flowers*, by F. E. Hulme, F.L.S.; 200 coloured plates and descriptive text; 5 vols., post 8vo. 62s. 6d. Cassell and Co.

Carefully prepared, with artistic representations of the plants described. It is a work in which scientific difficulties are avoided, and additional volumes would extend its usefulness.

4. *The Flowering Plants, Sedges, Grasses, and Ferns of Great Britain, with their Allies*, by Anne Pratt; coloured illustrations of 1,644 species; 6 vols., 8vo. 75s. Warne.

The Flowering Plants of Great Britain, by Anne Pratt; coloured illustrations of 1,340 species; 3 vols., 8vo. 42s. Warne.

The Ferns of Great Britain, with their Allies, by Anne Pratt; coloured illustrations of 63 species; 8vo. 12s. 6d. S.P.C.K.

Anne Pratt's works are too well known to require much comment or recommendation. The descriptions are not sufficient for a scientific identification of the plants, but illustrations of almost every British species are attached, and the popular portions are most interesting. The three publications mentioned above are similar. The first is the original and complete edition, or its re-issue; the second is an issue of the first five volumes, with the letterpress printed in smaller type, and the original plates retained. The third is the latter part of the sixth volume published separately. The Sedges and Grasses have not been so issued at present.

Wild Flowers, by Anne Pratt; with 192 coloured plates; 2 vols., 16mo. 12s. S.P.C.K.

This is a simple guide to the flowers of the fields and hedges, but it has in its day given to many, as to the writer himself, the first introduction to a friendship with Nature which will stand true for ever.

SCIENTIFIC TREATISES.

Amongst those held in most general estimation at the present time are :—

5. *The Student's Flora of the British Islands*, by Sir J. D. Hooker; 3rd edition; post 8vo. 10s. 6d. Macmillan.

This is the latest authority on English botany : to a critical description of every plant, the geographical range is also added. There are no illustrations in this work, but it is by far the most useful and interesting to the student.

6. *A Manual of British Botany*, by Prof. C. C. Babington; 12mo. 10s. 6d. Van Voorst. A thin paper edition for field use, 12s. 6d.

A work of long-recognised merit. There are no illustrations, but the descriptions are valuable, and special points are emphasised to facilitate identification.

7. *A Handbook of the British Flora*, by George Bentham; 1,295 wood-cuts; 2 vols., post 8vo. £1 1s. Reeve.

This handbook has found many admirers. It departs, however, so widely from the present accepted definition of species and varieties that confusion is inevitable.

The plates which accompany it in a second volume are carefully executed, but are too minute to insure safe guidance.

8. *The Botanist's Pocket Book*, by W. R. Hayward; crown 8vo. 4s. 6d. George Bell.

This is only a small key of genera and species for field use. The writer has used it constantly with doubtful satisfaction.

9. *Sowerby's English Botany*, containing a description and life-size drawings of British plants, edited by Boswell Syme; 1,923 coloured plates; 12 vols., imp. 8vo. £24 3s. George Bell. (The 12th volume, containing Ferns and Allies, with General Index, 35s. cloth.)

This is the standard work on British Botany. It has seen many alterations and additions since it was first issued in 1790. An edition in 12 volumes, published in 1832—1846, with descriptions by Sir James Smith, is sometimes met with ; it is carefully executed, and the plates of Flowering and Cryptogamic Plants not to be surpassed ; but, as it is arranged on the Linnean system, it is somewhat out of date. A good copy is worth about £12. The present and third edition (1863—1886) is an entire revision and re-arrangement to suit the Natural Order system and the extended knowledge of the British Flora; it only contains the Flowering Plants, Ferns and allies.

For many the possession of all the above-named works is needless, perhaps impossible, though it may be worth remembering that many booksellers (Edward Bumpus, Holborn Bars, E.C., for instance) will allow 25 per cent. off the prices here quoted. For practical purposes, however, sufficient would be found in Nos. 1 and 5, or, better still, in Nos. 4 (6 vols.) and 5; perhaps in No. 7 alone, used with caution.

To these should be added a book on Structural Botany: Oliver's *Lessons in Elementary Botany* (Macmillan, 4s. 6d.) is as good as any. There is a work—recently published—by F. A. Messer (10s. 6d.), which is a praiseworthy attempt to resist the unsatisfactory and unworthy plan of identifying plants by pictures, without a knowledge of their structural peculiarities. Sections are given of the critical parts of the plants, and if the principle were extended to the species as well as the genera, a want would be supplied and an evil checked. A most interesting book is Le Maout and Decaisnes' *General System of Botany, Descriptive and Analytical*, translated by Mrs. Hooker, with 5,500 figures and sections (31s. 6d., Longmans). It embraces exotic as well as British orders. The chief authority on the whole subject is Sach's *Text Book of Botany*, translated by Bennett and Dyer (31s. 6d., Clarendon Press), in which the organism of plants is subjected to a rigid and critical analysis.

Having spoken of the aids to a knowledge of Botany, and intending to append a few hints to collectors, a protest must here be made against the pernicious custom of rooting up wild plants for transference to the garden. The inevitable result must be a complete annihilation of all interesting species, as there are but few gardens where their natural surroundings can be supplied and their growth insured. From the first the Club has endeavoured to discountenance this practice, with what success the total destruction of the rare Brittle Bladder Fern in its new locality will show; if that fern is extinct in Harleston, it is probably lost to the county. Foreseeing, then, that the publication of the localities of the flowering plants of Harleston might be their death-knell, the writer has abstained from giving much of the detailed information possessed, preferring to direct those who desire to obtain specimens without injury to the living plant to the members of the Club, on whose authority the records are given. It is a truly selfish principle which robs the woods and hedges of flowers given for the enjoyment of those who, as the writer himself, have no other flower garden, and ask no other. Nor can it be for one moment supposed that our modest English flowers will yield the true pleasure for which they were created when they are ranked side by side, in seeming mockery, with the gay productions of foreign countries.

Trusting that in the student of Nature love and reverence will go hand in hand, the writer ventures to give the following few suggestions on the collection and preservation of plants based on his own experience:—The chief requisite for collecting is a long and narrow tin-box, in which the specimens may be placed when gathered, and in which they will keep fresh for some hours. To insure *lasting* specimens for the herbarium the plant should not be gathered when soaked with rain or heavy dew; if such is unavoidable, the moisture must be damped off with blotting-paper before pressure is applied. For drying, use a rough paper *without glaze*, and fairly thick. Some kinds of newspaper, blotting-paper (if often changed), and an absorbent brown paper used by grocers and sold sometimes as botanical drying-paper are good for the purpose. The plant must be placed between the sheets as evenly as possible; if the stem is thick, it is advisable to take a slice off one side; and if it prevents the pressure resting on the petals—as in the *Rosæ* and *Rubi*, for instance—a pad of blotting-paper under the flower will prevent shrinkage. A perfect specimen should contain flower and fruit; if the root is a *critical* part, it should also be added. The plant, thus prepared and placed in the drying-paper, must be put between boards under a *strong* and *even* pressure. After the second or third day it should be examined; as it will then be less rigid, the leaflets and petals may easily be set out. Pressure even stronger than before must again be applied, and for succulent species the paper changed occasionally.

When completely dry the plant should be mounted on stiff white paper. The size depends on the purpose of the collection; paper 17 in. by 10 in. will be found useful and workable, and can be obtained of most printers. It is a bad system to fasten the specimens on with glue or gum, as it renders it impossible to shift the mount, and the finer parts of the flowers are destroyed. The writer has found that very thin strips of parchment, cut with wider ends, laid across the stoutest parts of the plant and fastened to the paper with strong cement, form a most easy and effectual way of mounting with the least possible unsightliness.

To the paper must be attached the name (Latin and English), the Order, the date of collection, and the *locality* of the plant. Each Species should be placed, with its fellows of the same Genus, in a stout cover of blue or brown paper, bearing the generic name in the right-hand lower corner; the Genera in their turn should be included in another cover, bearing the name of the Natural Order to which they belong. The Orders may then be arranged after some such recognised list as the *London Catalogue* (Bell and Sons, price 6d.), and placed in a box or cabinet having many shelves to prevent undue pressure.

A systematic arrangement, insuring easy consultation, is necessary if the collection is intended for practical reference, and not for a melancholy spectacle of faded beauty; for it is impossible to prevent certain colours from changing, though careful selection and drying will do much to save disappointment, and to maintain at least some traces of Nature's loveliness.

III.—AUTHORITIES, ABBREVIATIONS, AND SIGNS.

AUTHORITIES.

HARLESTON BOTANICAL CLUB.
1884–1887.

ABBREV.
- A. Buckingham, Herbert, M.R.C.V.S., Harleston.
- B. Candler, William, Harleston.
- C. Candler, Charles, Harleston.
- D. Cordwell, Walter R., Harleston.
- E. Donnison, Miss A. Stote, The Dove House, Mendham, Norfolk.
- F. Galpin, Rev. F. W., (late of) Harleston.
- G. Prentice, John G., Harleston.
- H. Wilson, Edward, Harleston.
- I. Cartwright, Miss Ethel, and Miss Frieda Guthe, (late of) Flixton, Suffolk.
- K. Candler, Edmund, Harleston.
- L. Owles, Frederick R., Harleston.
- M. Cann, Archibald, Harleston.
- N. White, Miss Mary de Lacy, Weybread Vicarage, Suffolk.
- O. Hanmer, Miss Alice, Weybread, Suffolk.

PRINTED RECORDS.

- (FB) *Flora Britannica*, by Sir J. E. Smith, 1800–1804.
- (BG) *The Botanist's Guide through England and Wales*, by Dillwyn and Turner, 1805.
- (WA) Withering's *Arrangement of British Plants*. The sixth edition (1818) was principally consulted.
- (NBG) *The New Botanist's Guide*, by H. Cottrell Watson, 1835–1837.
- (HS) Henslow and Skepper, *Flora of Suffolk*, 1860.
- (T) Trimmer, *Flora of Norfolk*, 1866.
- (TS) Trimmer, *Supplement to the Flora of Norfolk*, 1884.

MANUSCRIPT RECORDS.

(EAH) A Catalogue of Plants found in the neighbourhood of St. Margaret's, South Elmham, in Suffolk, and Brockdish in Norfolk, by the Rev. E. A. Holmes, M.A., F.L.S., 1833—1885.

(DC) A Catalogue of Plants found in the parish of Dickleburgh, 1860—1870.

(JM) A Catalogue of Plants found in the parish of Shimpling by the Rev. J. W. Millard.

(JC) A Catalogue of Plants found in the parishes of Hoxne, Billingford, Scole, Wacton, &c., by Mr. J. C. Collins, of Diss.

(JH) Specimens in the Herbarium of Miss Jeffes (Needham Market); communicated by the Rev. W. M. Hind, LL.D.

CONTRIBUTORS.

The Rev. J. Landey Brown, Norwich.
Mr. Samuel Carman, Harleston.
The Rev. C. T. Cruttwell, Denton Rectory.
The Rev. Spencer Fellows, Pulham Rectory.
Mr. Flint, Gawdy Hall, Harleston.
The Rev. H. Temple Frere, Burston Rectory.
Mrs. Hanbury Frere, Horham Rectory.
The Rev. W. M. Hind, LL.D., Honington Rectory, Bury.
The Rev. E. F. Linton, Sprowston Rectory.
Mrs. J. Sancroft Holmes, Gawdy Hall.
Mr. James Muskett (the late), Harleston.
Mrs. Pemberton, Denton House.
Miss Perowne, Redenhall Rectory.
Mr. F. Spalding, Colchester.
Mr. W. Squires, Harleston.

The compiler here takes the opportunity of thanking the above-named contributors for the information and help they have given; also Mr. H. D. Geldart of Norwich, Mr. Arthur Bennett, F.L.S., of Croydon, and Mr. Bagnall, A.L.S., of Aston, for their assistance in determining doubtful species; and Sir Hugh Adair, Bart., and J. Sancroft Holmes, Esq., for permission to explore the woods on the Flixton and Gawdy Hall estates.

SIGNS.

* * preceding the name of a plant denotes that it has been introduced, but is now established.
* † preceding the name of a plant denotes that it is considered as probably extinct, not having been observed for many years.
* etc. attached to the list of localities implies that it is not considered to be exhaustive.
* Localities connected by a semi-colon and followed by an abbreviated name rest on the same authority.
* A bracketed abbreviation followed by an initial letter signifies that the record has been verified by a member of the Club during the years 1884—1887.
* The usual period of flowering is denoted by the number of the months following the English name. It can only be considered approximate.

---

* The nomenclature is that adopted in the *London Catalogue*, 8th edition, 1886. The synonyms of the 7th edition are added in brackets.
* Interleaved copies of the Flora are supplied for recording personal observations and ascertaining the distribution of plants in the district.

IV.

A LIST OF
THE FLOWERING PLANTS, FERNS
AND THEIR ALLIES.

FLOWERING PLANTS.

DICOTYLEDONES.

RANUNCULACEÆ.

CLEMATIS, L.

C. Vitalba, L. *Traveller's Joy.* 6—8. Frequent in hedges and thickets: Weybread Road, near the Heath House; Needham; Brockdish; Rushall; Dickleburgh; Long Stratton; Mendham; Wingfield; Denton; Alburgh; the Bath Hills, Ditchingham, etc. *Cf.* Introd., p. 27.

THALICTRUM, L.

T. flavum. L. *Meadow Rue.* 6, 7. Abundant by the sides of streams: The Waveney Marsh Dykes (Homersfield Bridge, etc.); Redenhall Beck; Flixton; Dickleburgh, etc.

ANEMONE, L.

A. nemorosa, L. *Wood Anemone.* 4, 5. Frequent in woods and groves: Homersfield; Harleston Wilderness Copse; Mendham Grove, Norfolk; Flixton; Dickleburgh, etc.

*A. apennina, L. *Blue Anemone.* 5, 6. Established for many years in a copse at Denton House (Mrs. Pemberton) F. *Cf.* Introd., p. 26.

ADONIS, L.

A. autumnalis, L. *Pheasant's Eye.* 5—8. Very rare: on land near Gawdy Hall (EAH). It occurs as a weed in gardens, F.

MYOSURUS, L.

+M. minimus, L. *Mouse-tail.* 4—6. Very rare. Sandy fields at Earsham (BG and WA), but has not been found for many years (NBG).

RANUNCULUS, L.

R. circinatus, Sibth. *Rigid-leaved Water Crowfoot.* 5—7. Common in ponds and dykes: the Waveney Marshes; Flixton Park; Denton, etc.

R. fluitans, Lam. *River Water Crowfoot.* 6, 7. Rare: in the Waveney below Syleham (EAH); introduced into the Redenhall Beck, F.

R. trichophyllus, Chaix. *Hair-leaved Water Crowfoot.* 6—8. Frequent in ditches and ponds: Mendham Marshes; Shotford; Rushall Wood; Dickleburgh; Shelton; St. Margaret's.

R. Drouetii, Godr. *Drouet's Water Crowfoot.* 5, 6. Abundant in streams and ponds: Moat at Ant Hill Farm, Redenhall; ponds, Harleston; Lush Bush; Needham; Weybread; plentiful in the Waveney at Earsham, F.

R. peltatus, Schrank. *Common Water Crowfoot.* 4—6. Pond on Harleston Common, F.

Var., **floribundus**. Very common in ponds and ditches, often growing with R. Drouetii.

R. sceleratus, L. *Celery-leaved Crowfoot.* 4—8. Abundant in the marsh dykes: Weybread, Needham, Mendham, etc. Pond-sides at Harleston, Flixton, etc.

R. Flammula, L. *Lesser Spearwort.* 6—8. Frequent in damp places: Gawdy Hall Wood; Needham Alder Carr; Fir Cover, Brockdish; Rushall Wood; Dickleburgh; Flixton.

R. Lingua, L. *Greater Spearwort.* 6—8. Rare: moist places, Hoxne (JC). Sides of the Waveney below the Bath Hills, Ditchingham (BG).

R. auricomus, L. *Wood Crowfoot.* 4—6. Common in bushy places: Cuckoo Hill, Mendham; Mendham Long Lane; Flixton, etc. Popular name *Goldilocks*.

R. acris, L. *Meadow Crowfoot.* 6—8. Common in meadows and by roadsides. Popular name *Butter-cup*.

Var., **vulgatus**. Abundant in the marshes.

R. repens, L. *Creeping Crowfoot.* 6—8. Common in pastures and meadows.

R. bulbosus, L. *Bulbous Crowfoot.* 5, 6. Common by roadsides and in meadows.

R. Sardous, Crantz. (R. hirsutus, Curtis). *Hairy Crowfoot.* 6—10. Rare: in a field in front of Weybread Lodge, D; St. Margaret's (EAH); Shelton (TS).

R. parviflorus, L. *Small-flowered Crowfoot.* 5—8. Very rare: at Harleston (NBG). Bedingham (TS).

R. arvensis, L. *Corn Crowfoot.* 6, 7. Frequent in cultivated fields: near Harleston Green Lane; Weybread Rifle Range; Rushall; Dickleburgh; Shimpling; Flixton, etc.

R. Ficaria, L. *Lesser Celandine.* 4, 5. Common in meadows and on shady banks.

CALTHA, L.

C. palustris, L. *Marsh Marigold.* 4—6. Common in the marshes of the Waveney and in meadows.

HELLEBORUS, L.

H. viridis, L. *Green Hellebore.* 3, 4. Very rare. Bushy places near Stradbrook (HS). Very poisonous.

H. fœtidus, L. *Fœtid Hellebore.* 3, 4. Rare: rather plentiful in lanes at St. Margaret's, and between St. Cross and Flixton (EAH) F. Bath Hills, Ditchingham (NBG and Mr. F. Spalding). Laxfield (BG). Very poisonous.

ERANTHIS, Salisb.

*E. hyemalis, Salisb. *Winter Aconite.* 2, 3. Gawdy Hall Great Wood, C. In Flixton Long Plantation (EAH) F. *Cf.* Introd., p. 26.

AQUILEGIA, L.

A. vulgaris, L. *Columbine.* 5—7. Rare: near St. Cross Church and at St. Margaret's (EAH) F. Hedgerows at Denton, B, and between Denton and Alburgh, H. Weybread (JH).

DELPHINIUM, L.

†D. Ajacis, Reich. *Branching Larkspur.* 6, 7. Cornfields at Earsham (BG), but not confirmed since. This is the D. Consolida, L., of earlier botanists.

ACONITUM, L.

†A. **Napellus,** L. *Monkshood.* 5—7. Formerly plentiful in a ditch at St. Peter's, but now lost owing to alteration of the locality (EAH). Very poisonous.

BERBERIDEÆ.

BERBERIS, L.

†B. **vulgaris,** L. *Barberry.* 5, 6. An old bush formerly in Gawdy Hall Wood (Mr. Flint).

NYMPHÆACEÆ.

NUPHAR, L.

N. **luteum,** Sm. *Yellow Water Lily.* 6, 7. Abundant in the Waveney; the Redenhall and the Weybread becks.

NYMPHÆA, L.

N. **alba,** L. *White Water Lily.* 7. Not infrequent in the Waveney (Syleham, Needham, Mendham, Flixton, etc.).

PAPAVERACEÆ.

PAPAVER, L.

*P. **somniferum,** L. *White Poppy.* 6, 7. Occasionally in cultivated ground and by way-sides, F.

P. **Rhœas,** L. *Common Red Poppy.* 6—8. Common on cultivated ground. Seed-vessel assuming the shape of a smooth round head.

P. **dubium,** L. *Smooth long-headed Poppy.* 6, 7. Frequent: Starston Railway Bridge; sand-pit on Withersdale Road. Mendham; Homersfield, etc. Apparently only var., **Lamottei,** (Lond. Cat., 7th ed.).

P. **Argemone,** L. *Rough long-headed Poppy.* 6, 7. Frequent in sandy places: Redenhall Road; Mendham; Dickleburgh; Earsham, etc.

P. **hybridum,** L. *Rough round-headed Poppy.* 5—7. Very rare. On an old wall at Dickleburgh (DC). I have found it in a similar situation in Dorset, F.

CHELIDONIUM, L.

C. majus, L. *Celandine.* 5—8. Frequent in hedges: Redenhall Road; Mendham; Weybread; Dickleburgh, etc.

FUMARIACEÆ.

CORYDALIS, DC.

*C. bulbosa, DC. (C. Solida, Hook.). *Tuberous Fumitory.* 5—7. A weed in shrubberies at Wortwell.

*C. lutea, DC. *Yellow Fumitory.* 5—8. On walls and waste ground at Harleston.

FUMARIA, L.

F. pallidiflora, Jord. (F. capreolata, L.) *Rampant Fumitory.* 5—8. Var., Boræi. Rare: in hedges near Weybread Church, F.

F. officinalis, L. *Common Fumitory.* 5—8. Common in fields and by waysides: London Road, Harleston; Needham; Brockdish; Dickleburgh; Wortwell; Flixton, etc.

CRUCIFERÆ.

CHEIRANTHUS, L.

*C. Cheiri, L. *Wall-Flower.* 4—6. On an old wall in Ellis' Yard, Harleston, formerly in great abundance, F. Plentiful on the ruins of Bungay Castle.

NASTURTIUM, R.Br.

N. officinale, R.Br. *Common Watercress.* 5—10. Common in streams and dykes.

N. sylvestre, R.Br. *Creeping Yellow-cress.* 6—9. Rare: in a dry brook below St. Cross Rectory (EAH), F. On Earsham Common (BG).

N. palustre, DC. *Marsh Yellow-cress.* 6—9. Common in wet places: Wortwell Marshes; Stow Fen, Earsham; Mendham; Harleston Common; Needham; Dickleburgh.

N. amphibium, R.Br. *Great Yellow-cress.* 6—9. Not common: sides of pond, Harleston Common, B. Pond in meadow adjoining Wilderness Lane, D. Syleham (EAH). Tivetshall, F. Dickleburgh (DC).

BARBAREA, R.Br.

B. vulgaris, R.Br. *Common Yellow Rocket.* 5—8. Common in meadows, fields, and by the sides of streams.

ARABIS, L.

A. perfoliata, Lam. (Turritis glabra, L.). *Smooth Tower Mustard.* 6—8. Not uncommon on dry banks: plentiful near Wortwell Schoolroom (1885); between Homersfield and St. Cross; below Homersfield Church (EAH), F. Near Wortwell Windmill (BG). Flixton (NBG). Scole (JC).

CARDAMINE, L.

C. amara, L. *Bitter Lady's Smock.* 4—6. Abundant in the meadows of the Waveney Valley (Weybread Water Mill, etc.).

C. pratensis, L. *Meadow Lady's Smock.* 4—6. Common in meadows and moist places: occasionally a double form is found. Popular name *Cuckoo Flower.*

C. hirsuta, L. *Hairy Lady's Smock.* 4, 5. Common on dry banks and walls.

C. flexuosa, With. (C. sylvatica, Link.). *Creeping Lady's Smock.* 4—6. Rare: in shady places near Wortwell Water Mill (EAH).

EROPHILA, DC.

E. vulgaris, DC. (Draba verna, L.). *Common Whitlow Grass.* 3—5. Very common on dry banks. One of the earliest wayside flowers.

COCHLEARIA, L.

*C. Armoracia, L. (Armoracia rusticana, BM). *Horse Radish.* 5—8. Abundant in a ditch at Flixton Village, F. Mendham, A.

HESPERIS, L.

*H. matronalis, L. *Dame's Gilliflower.* 5—7. Rare: in a cultivated field on Balking Hill, Harleston, H.

SISYMBRIUM, L.

S. Thaliana, Hook. (Arabis Thaliana, L.). *Thale Cress.* 4—7. Abundant on dry banks.

S. officinale, Scop. *Yellow Hedge Mustard.* 6—8. Common in hedge-banks and by roadsides.

S. Sophia, L. *Flixweed.* 6—8. Not uncommon in waste places: Mendham Old Priory, F. Shotford Heath, D. Homersfield Village, K.

S. Alliaria, Scop. (**Erysimum Alliaria, L.**). *Garlic Mustard.* 5, 6. Common in hedge-banks and woods.

ERYSIMUM, L.

E. Cheiranthoides, L. *Wallflower Mustard.* 6—8. Abundant in cultivated fields.

BRASSICA, L.

*B. Napus, L. *Rape.* 5—7. Banks and borders of fields: Harleston, etc.

*B. Rapa, L. *Turnip.* 5, 6. Waste places and borders of fields: Redenhall, etc.

*B. nigra, Koch. (**Sinapis nigra, L.**). *Black Mustard.* 6—8. Not common: sides of the stream between Harleston and Redenhall (T), and also at Wortwell, F. This is generally supposed to be the Mustard Plant of the Bible.

B. Sinapis, Vis. (**Sinapis arvensis, L.**). *Wild Mustard.* 5—8. A common weed in cultivated ground. Popular name *Charlock.*

B. alba, Bois. (**Sinapis alba, L.**). *White Mustard.* 6—8. Probably frequent, but not satisfactorily determined. Cultivated in gardens.

DIPLOTAXIS, DC.

D. muralis, DC. *Sand Rocket.* 6—9. Rare: roadside, Earsham Village, F; Shelton, Tivetshall (T). Bungay, D.

CAPSELLA, DC.

C. Bursa-pastoris, DC. *Shepherd's Purse.* 3—9. Everywhere, and of various forms.

SENEBIERA, DC.

S. Coronopus, Poir. (**Coronopus Ruellii, Gært.**). *Common Wart Cress.* 6—9. Common in waste places: the Boys' School, Harleston; St. Margaret's, etc.

LEPIDIUM, L.

L. **campestre**, R.Br. *Common Field Pepperwort.* 5—8. Frequent in fields and waste places: Baker's Barn Brickyard; Shotford; Rushall; Dickleburgh; Earsham, etc.

L. **Smithii**, Hook. *Hairy Field Pepperwort.* 6—8. Rare: hedge-banks, Flixton (EAH).

THLASPI, L.

T. **arvense**, L. *Penny Cress.* 5—7. Common in cultivated ground: Redenhall; Wortwell; Weybread; Pulham; Shimpling, etc.

TEESDALIA, R.Br.

T. **nudicaulis**, R.Br. *Naked-stalked Teesdalia.* 4—6. Rare: Homersfield Heath, F; Needham Sandpit (BG); the Bath Hills, Ditchingham (NBG). *Cf.* Introd., p. 31.

RAPHANUS, L.

R. **Raphanistrum**, L. *Wild Radish.* 6—9. Frequent in waste ground: railway cutting, Redenhall; Weybread Targets, etc.

RESEDACEÆ.

RESEDA, L.

R. **lutea**, L. *Wild Mignonette.* 6—8. Not common: Gatehouse gravel pit, Redenhall, D. Earsham, F. Scole, Billingford (JC). Formerly at Dickleburgh (DC).

R. **luteola**, L. *Dyer's Weld.* 6—8. Frequent in hedgebanks and waste places: near Harleston Station; below Homersfield Church; Earsham; Flixton; Scole; Billingford; Needham, etc.

VIOLARIEÆ.

VIOLA, L.

V. **palustris**, L. *Marsh Violet.* 5—7. Very rare: in the Spring Meadow, Dickleburgh (DC). Reported also from Flixton.

V. **odorata**, L. *Sweet Violet.* 2—5. Common in woods and hedge-banks. Var. **alba**, frequent: Mendham Road, etc.

V. sylvatica, Fr. *Wood Violet.* 3—6. Common in woods and hedge-banks. This is var. **Riviniana,** Reich. Popular names *Horse* or *Dog Violet.*

V. Reichenbachiana, Bor. *Lesser Wood Violet.* 3—6. Not common: Gawdy Hall Great Wood (Rev. E. F. Linton); Redenhall Lanes, F. St. Margaret's (EAH). *Cf.* Introd. p. 26.

V. arvensis, Murr. *Small-flowered Field Pansy.* 4—10. Common in cultivated ground. Popular name *Heart's-ease.*

POLYGALEÆ.
POLYGALA, L.

P. vulgaris, L. *Common Milkwort.* 5—9. Not common: near Capt. Moore's Farm, Needham; Homersfield Heath; Dickleburgh.

CARYOPHYLLEÆ.
DIANTHUS, L.

D. Armeria, L. *Deptford Pink.* 7, 8. Occasional: near Harleston (NBG); Balking Hill, E. Well's Lane, D. Mendham (EAH), B. Near the White House, Harleston; roadside opposite Middleton Hall, Mendham, G. Between Denton and Earsham (Mrs. Pemberton).

SAPONARIA, L.

S. officinalis, L. *Common Soapwort.* 7, 8. Rare: in hedges near houses; between Flixton and Bungay (BG), F. Formerly in a hedge-bank on the Needham Road near the first milestone (Mr. Samuel Carman).

SILENE, L.

S. Cucubalus, Wib. (S. inflata, Sm.). *Bladder Campion.* 6—8. Common in fields and by roadsides.

Var., **puberula**; Earsham (T).

S. gallica, var. **anglica,** L. *English Catch-fly.* 6—8. A weed in the Rectory Garden, St. Margaret's (EAH).

S. nutans, L. *Nottingham Catch-fly.* 6—8. Borders of fields: several plants a few years ago on a wild bank near the Little Barn, Gawdy Hall North Lodge (Mrs. J. Sancroft Holmes), F. It is not cultivated in gardens, and the locality appears natural. *Cf.* Introd., p. 26.

S. noctiflora, L. *Night-flowering Catch-fly.* 7, 8. Frequent on the clay: cultivated fields, Dickleburgh, Shimpling, F; St. Margaret's (EAH). Earsham; Shelton (T).

LYCHNIS, L.

L. alba, Mill. (L. vespertina, Sibth.). *White Campion.* 6—10. Common in cultivated fields.

L. diurna, Sibth. *Red Campion.* 5—9. Common in bushy places and hedge-banks. Popular name *Robin Hood.*

L. Flos-cuculi, L. *Meadow Campion.* 5—8. Common in meadows and damp places. Popular name *Ragged Robin.* This is the true Cuckoo Flower, as its name implies.

L. Githago, Lam. (Agrostemma Githago, L.). *Corn Cockle.* 6—8. Abundant in corn-fields.

HOLOSTEUM, L.

H. umbellatum, L. *Umbelliferous Chickweed.* 3, 4. Very rare: sparingly on the ruins of Hoxne Abbey (JC, 1883). It has long been recorded for the neighbouring parish of Eye, but is rapidly disappearing from Norfolk and Suffolk, its only British habitats.

CERASTIUM, L.

C. quaternellum, Fenzl. (Mœnchia erecta, Sm.). *Upright Mouse-ear Chickweed.* 5—8. Rare: in a dry pasture adjoining Homersfield Heath, F.

C. semidecandrum, L. *Little Mouse-ear Chickweed.* 4, 5. Frequent on dry banks: Well's Lane, Harleston; below Homersfield Church; Dickleburgh.

C. glomeratum, Thuill. *Broad-leaved Mouse-ear Chickweed.* 3—9. Common on dry banks. Included under C. vulgatum, L.

C. triviale, Link. (C. viscosum, Sm.). *Narrow-leaved Mouse-ear Chickweed.* 4—10. Common on dry banks and waste places. Included under C. vulgatum, L.

STELLARIA, L.

S. aquatica, Scop. (**Cerastium aquaticum**, L.). *Water Chickweed.* 7—9. Frequent in moist places: ditches of the Waveney (Syleham, Mendham, Wortwell, Flixton, Earsham, etc.).

S. media, L. *Common Chickweed.* 3—10. Common in waste places and cultivated ground.

S. Holostea, L. *Greater Stitchwort.* 4—6. Common in hedgebanks and bushy places.

S. palustris, Ehrh. (**S. glauca**, With.). *Marsh Stitchwort.* 5—7. Rare: in ditches of the Waveney, Billingford (JC), and about Bungay (NBG).

S. graminea, L. *Lesser Stitchwort.* 5—8. Frequent in bushy places: Well's Lane, Harleston; Flixton; St. Margaret's, etc.

S. uliginosa, Murr. *Bog Stitchwort.* 5, 6. Common in marshy places: Gawdy Hall Wood; Weybread, etc.

ARENARIA, L.

A. trinervis, L. *Three-nerved Sandwort.* 5—8. Frequent in shady places: Shotford Hill; Starston, Weybread, Flixton, etc.

A. serpyllifolia, L. *Thyme-leaved Sandwort.* 5—8. Common in dry waste places.

Var. **leptoclados**, Guss. Frequent: Needham Alder Carr Pit; Flixton, etc.

SAGINA, L.

S. apetala, L. *Small-flowered Pearlwort.* 5—9. Common on walls: Harleston; Redenhall; Pulham Market, etc.

S. procumbens, L. *Creeping Pearlwort.* 5—9. Common on walls and in waste places.

SPERGULA, L.

S. arvensis, L. *Corn Spurrey.* 6—8. Frequent in fields and waste places: near the White House, Harleston; Mendham Pit on Withersdale Road; Flixton.

LEPIGONUM, Fr.

L. rubrum, Fr. (Spergularia rubra, Fenzl.). *Field Sandwort Spurrey.* 5—8. Rare gravelly places in the neighbourhood of St. Margaret's (EAH).

HYPERICINEÆ.

HYPERICUM, L.

*H. calycinum, L. *Large-flowered St. John's Wort.* 7, 8. Mendham Grove, Norfolk, in the last century a garden. *Cf.* Introd., p. 29, note.

H. perforatum, L. *Common St. John's Wort.* 7, 8. Common in woods and hedge-banks.

H. quadratum, Stokes. *Square - stalked St. John's Wort.* 7. Frequent in moist places: Shotford; Brockdish; Dickleburgh; Earsham, etc.

H. humifusum, L. *Trailing St. John's Wort.* 7, 8. Rare: wood near Brockdish Hall (EAH). Weybread (JH). On Stuston Common (WA).

H. pulchrum, L. *Small Upright St. John's Wort.* 6, 7. Rare: waste ground opposite Hulk's Graves, Weybread, C. Mendham Priory Plantations, F. Hedenham (T).

H. hirsutum, L. *Hairy St. John's Wort.* 6—8. Not uncommon in woods: Gawdy Hall; Mendham; Dickleburgh; Flixton; St. Margaret's, etc.

†H. montanum, L. *Mountain St. John's Wort.* 7, 8. Rare: bushy places, Bath Hills (WA).

MALVACEÆ.

MALVA, L.

M. moschata, L. *Musk Mallow.* 7, 8. Frequent: Starston Road, near the railway bridge; Weybread; Needham; St. Margaret's.

M. sylvestris, L. *Common Mallow.* 6—9. Abundant in waste places.

M. rotundifolia, L. *Dwarf Mallow.* 6—10. Common by roadsides: Harleston; Brockdish; Wortwell; Homersfield; St. Margaret's, etc.

TILIACEÆ.

TILIA, L.

*T. vulgaris, Hayne. (T. intermedia, DC). *Common Lime.* 7. Frequent in plantations and hedgerows. T. platyphyllos, Scop., has been introduced at Redenhall.

LINEÆ.

LINUM, L.

L. catharticum, L. *Purging Flax.* 6—9. Frequent in dry pastures: Harleston Green Lane; near Gawdy Hall Wood; Shotford; Dickleburgh; Denton; Flixton, etc.

L. angustifolium, Huds. *Narrow-leaved Flax.* 6—8. Rare: on the banks of the railway cutting at Redenhall, F.

†L. usitatissimum, L. *Common Flax.* 7. Dry fields: Shimpling and Fressingfield (formerly). No longer cultivated in the district.

GERANIACEÆ.

GERANIUM, L.

*G. striatum, L. *Pencilled Crane's Bill.* 5, 6. Established in plantations at Denton (Rev. C. T. Cruttwell).

*G. phæum, L. *Dusky Crane's Bill.* 5, 6. Rare: bushy place, called Pound Hole, near Shotford Hall, L. On the site of an old garden, Shimpling (JM), K. Orchard, St. Margaret's (EAH), I.

G. pratense, L. *Blue Meadow Crane's Bill.* 6—9. Very rare: waste ground, Oaklands, Redenhall (perhaps an escape) B. Hoxne (JC).

G. pyrenaicum, L. *Mountain Crane's Bill.* 6, 7. Very common on banks by roadsides at Harleston. Needham (T); St. Margaret's (EAH); Oakley (JC).

G. molle, L. *Soft Crane's Bill.* 4—8. Common in waste places and on banks. A white variety is frequent.

G. pusillum, Burm. *Small-flowered Crane's Bill.* 6—9. Frequent in situations similar to G. molle. Harleston;

Needham; Brockdish; Dickleburgh; Mendham; St. Margaret's; Earsham.

G. dissectum, L. *Jagged-leaved Crane's Bill.* 5—8. Common in hedges and on the dyke-banks.

G. columbinum, L. *Long-stalked Crane's Bill.* 6, 7. Frequent in hedges and on banks: pit between Wortwell Schoolroom and Low Street; Shotford Hill; Mendham; Starston; Pulham.

G. lucidum, L. *Shining Crane's Bill.* 5—8. Rare: a weed on rockwork at Alburgh, C. Hedge-bank near St. Margaret's Church, Ilketshall (EAH).

G. Robertianum, L. *Herb Robert.* 4—9. Common in hedges and woods and on walls. With white flowers at Shimpling (JM).

ERODIUM, L.

E. cicutarium, Sm. *Common Stork's Bill.* 6—9. Common on sandy banks. A prolific flowerer. *Cf.* Introd., p. 31.

OXALIS, L.

O. Acetosella, L. *Wood Sorrel.* 5, 6. Not common: Gawdy Hall Wood (Miss Perowne), C. Mendham Priory Plantations, D. Billingford (JC). Fur Green Lane at Rushall (Mr. Arnold).

*__**O. corniculata,** L. *Procumbent Yellow Sorrel.* 6—9. Waste ground by roadside, Harleston.

ILICINEÆ.
ILEX, L.

I. Aquifolium, L. *Holly.* 5, 6. Frequent in hedgerows and woods: Redenhall; Starston; Shotford, etc.

CELASTRINEÆ.
EUONYMUS, L.

E. Europæus, L. *Spindle Tree.* 5—7. Not uncommon in hedgerows: Gawdy Hall; Shotford Hill; Dickleburgh; Shimpling; Flixton, etc.

RHAMNEÆ.

RHAMNUS, L.

R. catharticus, L. *Common Buckthorn.* 5—7. Not common: a bush on South Elmham Minster ruins; in the Rectory Paddock hedge, and occasionally elsewhere at St. Margaret's (EAH), F. Hedge near Weybread House. Hoxne (BG).

R. Frangula, L. *Alder Buckthorn.* 4—6. Rare: in plantations near Bungay (BG). Wacton (T).

SAPINDACEÆ.

ACER, L.

*A. pseudo-platanus, L. *Sycamore.* 5, 6. Frequent: Weybread; Starston, etc.

A. campestre, L. *Maple.* 5, 6. Common in woods and hedges. Fine trees at South Elmham Hall.

LEGUMINOSÆ.

GENISTA, L.

G. anglica, L. *Needle Green Weed.* 5, 6. Rare: sparingly on Wacton Common (TS and JC). Stuston (JC). Bungay Common (WA).

G. tinctoria, L. *Dyer's Green Weed.* 7, 8. Frequent: Wacton Common; Baker's Barn Brickyard, Redenhall; Weybread; Needham; Dickleburgh, etc.

ULEX, L.

U. Europæus, L. *Common Furze.* 3—6. Common in sandy and gravelly places. Popular name *Gorse*.

CYTISUS, Link.

C. Scoparius, Link. (Sarothamnus scoparius, Koch.). *Common Broom.* 4—6. Common, especially in gravel pits.

ONONIS, L.

O. repens, L. (O. arvensis, L.). *Procumbent Rest-Harrow.* 6—9. Frequent in barren pastures: Well's Lane, Harleston; Wortwell; Dickleburgh; Flixton.

O. spinosa, L. *Upright Rest-Harrow.* 6—9. Common in waste places and pastures. With white flowers at Dickleburgh (DC).

MEDICAGO, L.

*M. sativa, L. *Lucerne.* 6, 7. Near the top of Stubbing's Lane, Weybread, D.

M. lupulina, L. *Black Medick.* 5—8. Common in waste ground and pastures. Popular name *Black Nonsuch.*

MELILOTUS, Lam.

M. altissima, Thuill. (M. officinalis, Willd.). *Common Melilot.* 6—8. Not uncommon: fields near Harleston Bridge; Gawdy Hall; Mendham; Weybread; St. Margaret's; Flixton; Dickleburgh; Scole; Shimpling.

TRIFOLIUM, L.

T. subterraneum, L. *Subterranean Clover.* 5—8. Rare: in a grassy pit of footpath-field near Starston Rectory, C; near Harleston Bridge, D. St. Margaret's (EAH).

T. pratense, L. *Red Clover.* 5—9. Common in meadows and by roadsides.

T. medium, L. *Zigzag Clover.* 5—9. Rare: Stubbing's Lane, Weybread, D; Flixton, I; St. Margaret's (EAH); Bath Hills (WA).

T. ochroleucum, L. *Sulphur-coloured Clover.* 6—8. Very frequent in pastures: Mendham Long Lane; Homersfield; St. Margaret's; Baker's Barn, Redenhall; Thorpe Abbots; Dickleburgh; Shelton; Denton; Bath Hills, etc. *Cf.* Introd., p. 30.

*T. incarnatum, L. *Crimson Clover.* 6, 7. Roadsides, Needham, Billingford, Shimpling.

T. arvense, L. *Hare's-foot Clover.* 7, 8. Frequent in dry places: Needham Alder Carr Pit; Mendham Pit, on Withersdale Road; Homersfield; Dickleburgh, etc.

T. striatum, L. *Soft knotted Clover.* 6, 7. Common in dry pastures and gravel pits.

T. scabrum, L. *Rough rigid Clover.* 5—7. Rare: Needham Alder Carr Pit, F. Balking Hill, Harleston (BG); Bath Hills (WA).

T. glomeratum, L. *Smooth round-headed Clover.* 5, 6. Not infrequent in dry places: below Homersfield Church (EAH); gravel pit near Earsham Station, F. Bath Hills (FB).

*T. hybridum, L. *Alsike Clover.* 6—8. In fields and by roadsides: Baker's Barn Brickyard, Redenhall; near Mendham Bridge; pasture opposite Hulk's Graves, Weybread; D. *Cf.* Introd., p. 26.

T. repens, L. *White Clover.* 4—9. Common in pastures. This is the *Shamrock* of Ireland.

T. fragiferum, L. *Strawberry-headed Clover.* 7, 8. Frequent in meadows and pastures: Harleston Green Lane; pondside, Harleston Common: near the Moat, Gawdy Hall; Mendham; Weybread; Needham.

T. procumbens, L. *Hop Clover.* 6—8. Common in gravelly places.

T. dubium, Sibth. (T. minus, Sm.). *Lesser Hop Clover.* 5—7. Common in dry pastures.

T. filiforme, L. *Slender Clover.* 6, 7. Frequent: roadside near Denton House; grassy bank, Shotford Hill; dry pasture near Homersfield Heath, F. St. Margaret's (EAH), I.

ANTHYLLIS, L.

*A. vulneraria, L. *Common Kidney Vetch.* 5—8. In a grass field near Chediston, apparently introduced with seed, K.

LOTUS, L.

L. corniculatus, L. *Common Bird's-foot Trefoil.* 6—8. Common in meadows and pastures. Popular names *Ladies' Slippers, Shoes and Stockings.*

L. pilosus, Beeke. (L. major, Scop.). *Greater Bird's-foot Trefoil.* 6—8. Common in damp places.

ORNITHOPUS, L.

O. perpusillus, L. *Common Bird's-foot.* 5—7. Not infrequent on dry banks: Well's Lane, Harleston; Homersfield; pasture opposite Hulk's Graves, Weybread.

HIPPOCREPIS, L.

†H. comosa, L. *Horse-shoe Vetch.* 5—8. Very rare: pastures, Ditchingham (BG).

ONOBRYCHIS, Tourn.

*O. sativa, Lam. *Sainfoin.* 6, 7. Mendham Pit, on Withersdale Road; Alder Carr Pit, Needham; St. Margaret's. Probably escaped from cultivation.

VICIA, L.

V. hirsuta, Koch. *Hairy Tare.* 6—9. Common in dry bushy places and on banks: Needham Alder Carr Pit, etc

V. tetrasperma, Mœnch. *Four-seeded Slender Tare.* 6, 7. Frequent in hedges: The Hol-Way, Gawdy Hall: Weybread: Mendham Hill: Flixton, etc.

V. Cracca, L. *Tufted Blue Vetch.* 6—8. Frequent: Baker's Barn Brickyard, Redenhall; Needham Alder Carr Pit; Shotford Dykes; Dickleburgh, etc.

V. sepium, L. *Bush Vetch.* 5—7. Common in hedges and thickets. With white flowers near the White House, Harleston.

*V. sativa, L. *Common Cultivated Vetch.* 5—7. Borders of fields and waste ground: Harleston; Alburgh; Flixton etc.

V. angustifolia, Roth. *Common Wild Vetch.* 5—7. Waste places, especially on a sandy soil. Var. **segetalis**, common. Var. **Bobartii**, pit at Needham Hill, F.

V. lathyroides, L. *Spring Vetch.* 4, 5. Frequent in dry pastures: Shotford Heath; Needham Alder Carr Pit; Balking Hill, Harleston; Homersfield Pit; roadside below Homersfield Church ("apparently spreading," 1869, EAH). Bath Hills.

LATHYRUS, L.

†L. Aphaca, L. *Yellow Vetchling.* 5—8. Very rare: in a gravel pit with **Chlora perfoliata** at Ditchingham (NBG).

L. Nissolia, L. *Grass-leaved Vetchling.* 5, 6. Rare: by the side of the footpath from Harleston to Mendham (BG),

E; Mendham Hill (EAH). Gawdy Hall Great Wood (NBG and Rev. J. L. Brown). Thelveton Churchyard (WA).

L. pratensis, L. *Meadow Vetchling.* 6—8. Common in meadows and hedge-banks.

ROSACEÆ.

PRUNUS, L.

P. communis, Huds. (P. spinosa, L.). *Common Sloe.* 4, 5. Common in hedgerows. Popular name *Black-thorn.*

P. institia, L. *Bullace.* 4, 5. Occasionally in hedgerows: Lush Bush; Wortwell, etc.

*P. domestica, L. *Wild Plum.* 4, 5. Near the Heath House, Weybread; Flixton Village.

P. Avium, L. *Wild Cherry.* 5. Not common: by the footpath from Harleston Green Lane to Mendham, H. Weybread Village, O.

P. Cerasus, L. *Morella Cherry.* 5. Rare: Skeatsmere, Needham, D.

P. Padus, L. *Bird Cherry.* 5. Rare: Starston (introduced), F. Hoxne (JC). Bedingham, Earsham (BG).

SPIRÆA, L.

S. Ulmaria, L. *Meadow Sweet.* 6—8. Common in meadows and damp places.

RUBUS, L.

R. Idæus, L. *Common Raspberry.* 6, 7. Not common: Gawdy Hall Great Wood.

R. Lindleianus, Lees. *Lindley's Bramble.* Frequent: Gawdy Hall Great Wood: Mendham Priory Plantations, F. St. Margaret's (EAH).

R. rhamnifolius, W & N. *Buckthorn-leaved Bramble.* Rare: Flixton Long Plantation, F. *Cf.* Introd., p. 26.

R. incurvatus, Bab. *Curled-leaved Bramble.* Rare: Brockdish (EAH).

R. rusticanus, Merc. (R. discolor, W & N.). *Common Bramble.* 7, 8. Hedges everywhere, with many varieties.

R. leucostachys, Sm. *Long-clustered Bramble.* Frequent: Gawdy Hall Great Wood; Flixton Long Plantation and Abbey Wood, F.

R. Salteri, Bab. *Salter's Bramble.* Rare: Gawdy Hall Great Wood, F. *Cf.* Introd., p. 26.

R. carpinifolius, W & N. *Hornbeam-leaved Bramble.* Not common: St. Margaret's (EAH).

R. macrophyllus, W & N. *Large-leaved Bramble.* A variety is not uncommon in Gawdy Hall Wood and Mendham Priory Plantation, F.

R. scaber, W & N. *Rough-leaved Bramble.* Rare: Gawdy Hall Great Wood (Rev. E. F. Linton), F. *Cf.* Introd., p. 26.

R. Radula, W. *File-stemmed Bramble.* Frequent: Gawdy Hall Wood; Mendham Priory Plantations, F.

R. Koehleri, W. *Koehler's Bramble.* Var. **infestus**, bushy places opposite Hulk's Graves, Weybread, F.

Var. **pallidus**, St. Margaret's (EAH). *Cf.* Introd., p. 26.

R. Balfourianus, Blox. *Balfour's Bramble.* St. Margaret's (EAH).

R. corylifolius, Sm. *Hazel-leaved Bramble.* 6—8.

Var. **conjungens**, common. Var. **fasciculatus**, frequent: Lush Bush; Gawdy Hall Wood; Needham, F.

R. deltoideus, P.J.M. *Mallow-leaved Bramble.* Rare: St. Peter's Lane, South Elmham (EAH).

R. scabrosus, P.J.M. *Tubercular Bramble.* Rare: St. Margaret's (EAH).

R. cæsius, L. *Dewberry.* 6, 7. Common in damp places. Vars. **tenuis** and **hispidus**, Gawdy Hall Wood, F.

GEUM, L.

G. urbanum, L. *Common Avens.* 6—8. Common in hedgebanks and fields.

G. rivale, L. *Water Avens.* 6, 7. Rare: pondside, Stradbrooke Rectory Grounds (Mrs. Hanbury Frere). Moist places, Billingford (JC).

FRAGARIA, L.

F. vesca, L. *Wood Strawberry.* 4—7. Frequent on banks and in woods: Gawdy Hall; Denton; Flixton; Dickleburgh, etc.

*F. elatior, Ehrh. *Hautbois Strawberry.* 6—9. On roadside bank, Denton House Plantation, F.

POTENTILLA, L.

P. Fragariastrum, Ehrh. *Strawberry-leaved Cinquefoil.* 2—5. Common on dry banks: often confounded with **Fragaria vesca.**

P. Tormentilla, Neck. *Common Tormentil.* 5—8. Frequent in meadows and woods: Starston; Fir Cover, Brockdish; Mendham Priory Plantations; Flixton; St. Margaret's.

P. reptans, L. *Creeping Cinquefoil.* 6—9. Common on banks and by roadsides.

P. Anserina, L. *Silver-weed.* 5—7. Common by roadsides and in waste places.

P. argentea, L. *Hoary Cinquefoil.* 6—8. Frequent on sandy banks: Well's Lane, Harleston; Shotford Heath; Wortwell; Homersfield; Flixton; Earsham; Thorpe Abbots.

P. Comarum, Nestl. (Comarum palustre, L.). *Marsh Cinquefoil.* 5—7. Rare: ditches of the lane to Wortwell Cricket Field, F.

ALCHEMILLA, L.

A. arvensis, Lam. *Field Lady's Mantle.* 5—8. Common in cultivated fields and waste places.

AGRIMONIA, Tour.

A. Eupatoria, L. *Common Agrimony.* 6—8. Common on dry banks and by roadsides.

E

POTERIUM, L.

*P. muricatum, Spach. *Muricated Salad Burnet.* 5—8. Not infrequent on borders of cultivated fields: Needham Alder Carr Pit, D. Redenhall Gatehouse Pit, F. Mendham Hill, H. Near Starston Bridge, C. Generally with *Sainfoin*.

ROSA, L.

R. tomentosa, Sm. *Downy-leaved Rose.* 6, 7. Frequent: Gawdy Hall; Redenhall; St. Margaret's, etc.

   Var. subglobosa, Gawdy Hall Great Wood (Rev. E. F. Linton), F.

R. rubiginosa, L. *True Sweet-briar.* 6, 7. Frequent: lane near Harleston Station; Needham; Mendham; Abbey Wood, Flixton; Dickleburgh.

R. micrantha, Sm. *Small-flowered Sweet-briar.* 6, 7. Rare: hedges, St. Margaret's (EAH).

R. canina, L. *Common Dog Rose.* 6, 7. Vars. dumalis and urbica, common. Var. lutetiana, frequent: Harleston Green Lane, etc.

R. arvensis, Huds. *White-flowered Trailing Rose.* 6, 7. Frequent: Harleston Green Lane; Gawdy Hall Wood; Flixton, etc.

   Var. bibracteata, Gawdy Hall Wood.

PYRUS, L.

P. torminalis, Ehrh. *Wild Service Tree.* 5. Rare: woods, Bath Hills, Ditchingham (BG).

P. Aria, Sm. *Common White-beam.* 5. Rare: woods, Earsham (BG).

P. communis, L. *Wild Pear.* 4, 5. Not common: hedges, St. Margaret's and St. Peter's (EAH). Shimpling (JM).

P. Malus, L. *Wild Apple.* 5. Frequent: Vars., acerba and mitis growing together in hedges near Starston Bridge.

CRATÆGUS, L.

C. oxyacantha, L. *Hawthorn.* 5, 6. Hedgerows: Var., oxyacanthoides, Redenhall Road, near Lush Bush. Var., monogyna, common. Popular name *May.*

SAXIFRAGEÆ.

SAXIFRAGA, L.

S. tridactylites, L. *Rue-leaved Saxifrage.* 3—6. Rare: on old walls at Brockdish (EAH), and Mendham, F. Dickleburgh (DC).

S. granulata, L. *White Meadow Saxifrage.* 5, 6. Abundant in pastures and on banks: Harleston; Redenhall; Homersfield; Hoxne, etc.

CHRYSOSPLENIUM, L.

C. alternifolium, L. *Alternate-leaved Golden Saxifrage.* 4—6. Rare: Spring Wood, Weybread (BG), K. In a shady lane by the river at Needham (BG). Plentiful in Flixton Long Plantation, I.

PARNASSIA, L.

P. palustris, L. *Grass of Parnassus.* 8, 9. Very rare: on Kett's Fen, Shimpling (JM), K.

RIBES, L.

*R. Grossularia, L. *Gooseberry.* 4, 5. Frequent in hedges and woods: Starston; Homersfield; Flixton; Shimpling, etc.

*R. rubrum, L. *Red Currant.* 4, 5. Not common: Homersfield Wood; bank of the Weybread Beck: Shotford Dykes: Gawdy Hall Great Wood.

R. nigrum, L. *Black Currant.* 4, 5. Occasionally by streams: Needham Alder Carr; Weybread Beck; near Flixton Village; Shimpling.

CRASSULACEÆ.

TILLÆA, L.

T. muscosa, L. *Mossy Tillæa.* 4—8. Rare: gravel walks at Hoxne (BG): also near Bungay (Mr. D. Stock).

SEDUM, L.

S. Telephium, L. *Live-long Orpine.* 7, 8. Var., **purpurascens,** frequent: Foxburrows Plantation, Weybread, C, D. Plentiful in a gravel pit near Earsham Station, F Homersfield, K.

S. acre, L. *Yellow Stone-crop.* 6, 7. Common on roofs of houses, walls, and dry banks.

S. rupestre, Huds. *Rock Stone-crop.* 7, 8. Rare: roadside between Needham and Brockdish, D.

SEMPERVIVUM, L.

*****S. tectorum,** L. *Common House-leek.* 7. Frequent on roofs and walls.

HALORAGEÆ.

HIPPURIS, L.

H. vulgaris, L. *Common Mare's-tail.* 6, 7. Common in shallow dykes and ponds: Shotford Bridge, etc.

MYRIOPHYLLUM, L.

M. verticillatum, L. *Whorled Water-Milfoil.* 7, 8. Not uncommon: ditches near Lush Bush; pond near Rushall Wood; Brockdish; frequent in the Mendham Marshes.

M. spicatum, L. *Spiked Water-Milfoil.* 5—7. Frequent in the marsh dykes: Needham; Shotford; Mendham, etc.

CALLITRICHE, L.

C. vernalis, Koch. *Vernal Water Starwort.* 4—9. Common in ditches and ponds.

C. stagnalis, Scop. (C. platycarpa, Kutz.). *Large-fruited Water Starwort.* 5—9. Frequent in the Mendham Marshes; ponds at Pulham Mary and Gissing (T).

C. hamulata, Kutz. *Hooked Water Starwort.* 4—9. Needham Marshes (Rev. E. F. Linton), and Shimpling, F.

LYTHRARIEÆ.

LYTHRUM, L.

L. Salicaria, L. *Purple Loosestrife.* 6—9. Common by the sides of the Waveney and water-courses.

PEPLIS, L.

P. Portula, L. *Water Purslane.* 6—8. Rare moist places, Topcroft (TS). Near Bungay (Mr. D. Stock).

ONAGRARIEÆ.

EPILOBIUM, L.

E. hirsutum, L. *Great hairy Willow-herb.* 7, 8. Frequent in damp places and on river-sides: banks of the Waveney; Harleston Green Lane; Redenhall Beck; Gawdy Hall Wood, etc.

E. parviflorum, Sch. *Small-flowered hairy Willow-herb.* 7, 8. Common in watery places. Var. **rivulare** (subglabrous form) common in the marshes.

E. montanum, L. *Broad-leaved Willow-herb.* 6—8. Common in woods and shady places.

E. tetragonum, L. *Square-stalked Willow-herb.* 8, 9. Frequent by the side of ditches and in damp situations: The Wilderness Pond, Harleston; Gawdy Hall Great Wood; Needham Marshes, etc.

E. palustre, L. *Narrow-leaved Willow-herb.* 7, 8. Not common: marshy ground; Brockdish (EAH), Dickleburgh (DC).

ŒNOTHERA, L.

*Œ. biennis, L. *Common Evening Primrose.* 7—9. Railway banks at Homersfield.

CIRCÆA, L.

C. lutetiana, L. *Enchanter's Nightshade.* 6—8. Frequent in shady places: Gawdy Hall Great Wood; Denton Plantations; Flixton Woods; Bath Hills; Dickleburgh; Billingford, etc.

CUCURBITACEÆ.

BRYONIA, L.

B. dioica, L. *Red-berried Bryony.* 6—8. Frequent in hedgerows: near Harleston Bridge; Baker's Barn, Redenhall; near Brockdish School; Dickleburgh; Flixton; Earsham.

UMBELLIFERÆ.

HYDROCOTYLE, L.

H. vulgaris, L. *Marsh Pennywort.* 5—8. Not common: marshy ground near Wingfield Castle, C. Lane to Wortwell Cricket Field, F. On Dickleburgh Moor (DC).

SANICULA, L.

S. europæa, L. *Wood Sanicle.* 5—7. Common in woods: Gawdy Hall; Homersfield; Mendham; Flixton; Denton; Dickleburgh; Billingford.

CONIUM, L.

C. maculatum, L. *Common Hemlock.* 6—8. Frequent in bushy places: Shotford Hill; Redenhall Beck; Mendham; Weybread; Dickleburgh; St. Margaret's; Flixton, etc. Very poisonous: stem spotted with purple.

SMYRNIUM, L.

S. Olusatrum, L. *Common Alexanders.* 5, 6. Frequent in waste places: between Gawdy Hall and Lush Bush, C. Bungay Road, Flixton, I; near Homersfield Wood, F. Between Harleston and Scole, and at Flixton Village (NBG), F. *Cf.* Introd., p. 31.

BUPLEURUM, L.

B. rotundifolium, L. *Perfoliate Hare's-ear.* 6, 7. Rare: a weed in the Mill House Ground, Jay's Green, Harleston, F.

APIUM, L.

A. nodiflorum, Reich. (Helosciadium nodiflorum, Koch.). *Procumbent Water Parsnip.* 6—8. Common in the marsh dykes: Shotford, Mendham, Wortwell, Flixton, etc. On Dickleburgh Moor.

A. inundatum, Reich. (H. inundatum, Koch.). *Lesser Water Parsnip.* 6, 7. Wet places: not common: Wingfield, D. Dickleburgh Moor and Stow Fen, Earsham, F. Tivetshall (TS).

SISON, L.

S. Amomum, L. *Hedge Stonewort.* 7—9. Common in hedge-banks and by roadsides.

SIUM, L.

S. latifolium, L. *Great Water Parsnip.* 7, 8. Frequent by the sides of the Waveney (below Mendham Mill; Needham Alder Carr; Brockdish, etc). Dickleburgh.

S. erectum, Huds. (S. angustifolium, L.). *Upright Water Parsnip.* 7—9. Common in the marsh dykes and in streams.

ÆGOPODIUM, L.

Æ. Podagraria, L. *Gout-weed.* 6—8. Common in bushy and waste places.

PIMPINELLA, L.

P. Saxifraga, L. *Common Burnet Saxifrage.* 7—9. Common in meadows and on banks: plentiful on the steep bank of Redenhall Churchyard.

CONOPODIUM, Koch.

C. denudatum, Koch. (Bunium flexuosum, With.). *Earthnut.* 5, 6. Not common: meadows near Mendham Targets, C. Foxburrows Plantation, Weybread, D. Dickleburgh (DC).

CHÆROPHYLLUM, L.

C. temulum, L. *Rough Chervil.* 6—8. Common in fields and hedges: Harleston; Weybread; Mendham; Starston, etc. This and the poisonous *Hemlock* are the only British species of the Order having purple-spotted stems.

SCANDIX, L.

S. Pecten-Veneris, L. *Venus' Comb.* 5—9. Common in cultivated fields. Popular name *Shepherd's-needle.*

ANTHRISCUS, Pers.

A. vulgaris, Pers. (**Chærophyllum Anthriscus**, Lam.). *Common Chervil.* 5, 6. Frequent in hedge-banks; Wilderness Lane, Harleston; Starston Road; Flixton; St. Margaret's, etc.

A. sylvestris, Hoff. (**Chærophyllum sylvestre**, L.). *Wild Chervil.* 4—6. Common in hedge-banks and groves.

*A. Cerefolium, Hoff. (**Chærophyllum sativum**, Lam.). *Garden Chervil.* 6, 7. In great plenty on a bank near Halesworth, to all appearance wild (F.B., 1800). Growing quite wild on the roadside between Wisset and Halesworth (*id. loc.*, EAH).

FŒNICULUM, Hoff.

F. vulgare, Gaert. *Common Fennel.* 7, 8. Frequent: Wortwell; Harleston Bridge; Baker's Barn Brickyard, Redenhall; Starston; Shotford Heath; in a hedge on the Bath Hills, where it has grown for many years (NBG). *Cf.* Introd., p. 30.

ŒNANTHE, L.

Œ. fistulosa, L. *Common Water Dropwort.* 6—9. Frequent in the marsh dykes: Wortwell; Flixton; Mendham; Brockdish; Dickleburgh

Œ. Phellandrium, Lam. *Fine-leaved Water Dropwort.* 7—9. Frequent in ponds and slow ditches: near the White House, Harleston; Gawdy Hall Great Wood; near Starston Hall; Flixton; St. Margaret's; Dickleburgh, etc.

Œ. fluviatilis, Cole. *River Water Dropwort.* 6—9. Abundant in the Waveney (Syleham, Shotford, Homersfield, etc.). Also in the slow ditches adjacent, where it maintains its characters. First recorded for Norfolk in 1883.

ÆTHUSA, L.

Æ. Cynapium, L. *Common Fool's-Parsley.* 7, 8. Frequent in fields, gardens, and waste places: near Wortwell School; Mendham Long Lane; Starston Fields, etc. Somewhat like the *Parsley*, but very poisonous.

SILAUS, Bess.

S. **pratensis**, Bess. *Meadow Sulphurwort.* 6—9. Frequent in pastures and thickets: Harleston; Gawdy Hall; Redenhall; Mendham; Shimpling; Topcroft, etc.

ANGELICA, L.

A. **sylvestris**, L. *Wild Angelica.* 7—9. Frequent in moist woods: Gawdy Hall Great Wood; Rushall; Dickleburgh; St. Margaret's (EAH).

PEUCEDANUM, L.

P. **sativum**, Benth. (**Pastinaca sativa**, L.). *Wild Parsnip.* 6—8. Frequent, especially on clay soil: about Starston Hall and Baker's Barn, Redenhall; hedges, Jay's Green, Harleston, etc.

HERACLEUM, L.

H. **sphondylium**, L. *Common Cow-Parsnip.* 6—9. Common in hedges and fields.

DAUCUS, L.

D. **Carota**, L. *Wild Carrot.* 6—8. Frequent in fields and by waysides: Needham Hill; Mendham; Dickleburgh, etc.

CAUCALIS, L.

C. **arvensis**, Huds. (**Torilis infesta**, Spr.). *Field Hedge-Parsley.* 7, 8. Not common: fields near the Weybread Targets, D. St. Margaret's (EAH).

C. **Anthriscus**, Huds. (**Torilis Anthriscus**, Gaert.). *Upright Hedge-Parsley.* 6—9. Frequent in hedge-banks: Redenhall Road; Gawdy Hall Wood; Dickleburgh, etc.

C. **nodosa**, Scop. (**Torilis nodosa**, Gaert.), *Knotted Hedge-Parsley.* 6—9. Not uncommon on banks by waysides: Starston Road, near the Railway Bridge, H. Between Needham and Brockdish, F.

ARALIACEÆ.

HEDERA, L.

H. **Helix**, L. *Common Ivy.* 9—11. Common on trees and in woods and hedge-banks.

CORNACEÆ.

CORNUS, L.

C. sanguinea, L. *Common Dogwood.* 6—8. Frequent in copses and hedges: Harleston Green Lane; Wortwell; Flixton; Dickleburgh, etc.

CAPRIFOLIACEÆ.

ADOXA, L.

A. Moschatellina, L. *Tuberous Moschatel.* 4, 5. Frequent in woods and shady banks: Shotford Hill; Lush Bush; Redenhall; Homersfield Wood; Dickleburgh; Billingford.

SAMBUCUS, L.

S. nigra, L. *Common Elder.* 5, 6. Common in hedgerows and woods.

†S. Ebulus, L. *Danewort.* 6—8. Rare: waysides and waste places: Mendham Long Lane by Harleston (BG).

VIBURNUM, L.

V. Opulus, L. *Common Guelder-Rose.* 5—7. Frequent in damp situations: Weybread Beck; Needham Osier Ground; Gawdy Hall Wood; Rushall; Dickleburgh; Wacton; Scole; Flixton, etc.

V. Lantana, L. *Wayfaring Tree.* 5, 6. Rare: in Mendham Grove, Norfolk, K. *Cf.* Introd., p. 30.

LONICERA, L.

*L. Caprifolium, L. *Perfoliate Honeysuckle.* 5, 6. Not common, but well established: hedges near Needham Mill, Suffolk, F. Roadside between St. Margaret's and St. Peter's (EAH), F. Bath Hills (T).

L. Periclymenum, L. *Common Honeysuckle.* 6—9. Frequent in hedges and thickets: Harleston; Mendham; Flixton; Dickleburgh, etc.

RUBIACEÆ.

GALIUM, L.

G. cruciatum, With. *Crosswort.* 4—7. Frequent in hedge-banks and copses : Mendham Long Lane ; Well's Lane, Harleston ; near Shotford Hall ; Rushall Road ; Flixton ; St. Margaret's.

G. verum, L. *Yellow Bedstraw.* 6—9. Frequent : Redenhall Churchyard ; Needham Hill ; Shotford Hill ; Flixton ; Dickleburgh, etc.

G. erectum, Huds. *Narrow-leaved Great Bedstraw.* 6, 7. Rare : hedges, Brockdish (T).

G. Mollugo, L. *Great Bedstraw.* 6—9. Var. elatum, abundant : roadsides, Redenhall, Weybread, Needham ; gravel pits, Redenhall Gatehouse, Wortwell Broadwash, Flixton ; Dickleburgh, etc. A very strong growth when supported by bushes.

G. saxatile, L. *Heath Bedstraw.* 6—8. Dry pastures : not common : opposite Hulk's Graves, Weybread, D. Homersfield Heath, F.

G. palustre, L. *Marsh Bedstraw.* 7, 8. Common in marshy ditches and damp woods.

G. uliginosum, L. *Rough Marsh Bedstraw.* 7, 8. Not uncommon in damp places : between St. Margaret's and St. Cross ; Shotford Bridge, F. Dickleburgh (DC).

G. anglicum, Huds. *Wall Bedstraw.* 6—8. Very rare : on an old wall below Needham Mill, Norfolk, C.

G. Aparine, L. *Goose-grass.* 5—9. Common in bushy places. Popular name *Cleavers*.

G. tricorne, With. *Rough Corn Bedstraw.* 6—9. Very rare : field near cinder-path before entering Flixton Park, I. *Cf.* Introd., p. 30.

ASPERULA, L.

A. odorata, L. *Sweet Woodruff.* 5, 6. Rare : shady places, Dickleburgh (DC).

SHERARDIA, L.

S. arvensis, L. *Field Madder.* 4—10. Common in cultivated fields and on banks.

VALERIANEÆ.

VALERIANA, L.

V dioica, L. *Small Marsh Valerian.* 5, 6. Frequent in damp meadows : near Weybread Water Mill ; Brockdish ; Mendham ; Dickleburgh ; Flixton.

V. officinalis, L. *Great Wild Valerian.* 6—8. Common by riversides and in damp places.

CENTRANTHUS, DC.

*C. ruber, DC. *Red Valerian.* 6—9. In waste places and on walls : Harleston Common.

VALERIANELLA, Tour.

V. olitoria, Mœn. *Common Lamb's Lettuce.* 5—8. Common on banks and in gravel pits : below Balking Hill, Harleston ; Downs Farm, Homersfield ; Needham Alder Carr Pit, etc.

V. dentata, Poll. *Narrow-fruited Lamb's Lettuce.* 6—8. Not common : fields near Stubbings' Lane, Weybread ; Needham Green Lane ; near Rushall Road, D. Shimpling, F. Brockdish and St. Margaret's (EAH).

DIPSACEÆ.

DIPSACUS, L.

D. sylvestris, L. *Wild Teasel.* 7—9. Not uncommon in bushy places.

D. pilosus, L. *Small Teasel.* 7—9. Not uncommon : Gawdy Hall Great Wood ; Lush Bush ; Weybread Beck, F. Brockdish, D. St. Margaret's (EAH). Near Harleston (NBG) ; Homersfield (BG).

SCABIOSA, L.

S. succisa, L. *Devil's-bit Scabious.* 6—8. Not common: moist places: near Mendham Targets, C. Weybread Meadows, D. Shimpling, F. Dickleburgh (DC). Fressingfield, D.

S. columbaria, L. *Small Scabious.* 7, 8. Rare: dry banks: Bath Hills, Ditchingham (T). Dickleburgh (DC).

S. arvensis, L. (Knautia arvensis, Coult.). *Field Scabious.* 6—9. Common in fields and hedge-banks.

COMPOSITÆ.

EUPATORIUM, L.

E. cannabinum, L. *Hemp Agrimony.* 7—10. Frequent by the sides of streams; Shotford Bridge; Weybread Mill; Needham Mill; Brockdish; Dickleburgh, etc.

SOLIDAGO, L.

S. Virgaurea, L. *Common Golden-rod.* 7, 8. Rare: on waste ground opposite Hulk's Graves, Weybread (the late Mr. Muskett), E.

BELLIS, L.

B. perennis, L. *Common Daisy.* 3—9. Generally distributed.

ERIGERON, L.

E. acris, L. *Blue Fleabane.* 7—9. Rare: gravelly places, Needham Alder Carr Pit, E. Homersfield Heath (plentiful), F.

FILAGO, Tour.

F. germanica, L. *Common Cudweed.* 6—10. Frequent in cornfields and waste places: Shotford; Starston; Flixton, etc. Called by the old botanists *Herba impia*, because the young branches overtop the older head from which they spring.

F. minima, Fr. *Slender Cudweed.* 6—9. Not common: gravelly places: by the side of the Rectory House Wall, Redenhall; Flixton New Road, F. Dickleburgh (DC).

GNAPHALIUM, L.

G. uliginosum, L. *Marsh Cudweed.* 7, 8. Not common: pond-side near the Dove House, Mendham (Norfolk), E. Damp places between Shotford Bridge and Spurkett's Lane, D. Brockdish (EAH).

G. sylvaticum, L. *Upright Cudweed.* 7—9. Rare: bushy ground opposite Hulk's Graves, Weybread, E. Dickleburgh (DC).

INULA, L.

I. conyza, DC. *Ploughman's Spikenard.* 7—9. Frequent in hedge-banks and woods: Harleston Green Lane; Mendham Hill (Norfolk); Mendham High Road; Bath Hills, etc.

PULICARIA, Gaert.

P. dysenterica, Gaert. (Inula dysenterica, L.). *Greater Fleabane.* 7—10. Frequent in moist situations: Gawdy Hall Wood; Waveney Valley, etc.

BIDENS, L.

B. cernua, L. *Nodding Bur-Marigold.* 6—9. Frequent in the marsh dykes: Brockdish; Needham; Weybread; Wortwell; Dickleburgh Moor; Shimpling.

B. tripartita, L. *Tripartite Bur-Marigold.* 7—9. Rare: pond near Mendham Cross Roads, Norfolk; sides of lane to Wortwell Cricket Field, F.

ACHILLÆA, L.

A. Millefolium, L. *Common Yarrow.* 6—9. Common in banks by waysides.

A. Ptarmica, L. *Sneeze-wort Yarrow.* 7, 8. Rare: river-bank below Mendham Mill, H. Brockdish (EAH), D.

ANTHEMIS, L.

A. Cotula, L. *Stinking Mayweed.* 6—9. Frequent in waste places: Baker's Barn Brickyard, Redenhall; Mendham Pit, on Withersdale Road; St. Margaret's, etc.

A. arvensis, L. *Corn Chamomile.* 6—9. Rare: waste ground above the Alder Carr, Needham; near the Metfield footpath, Mendham, F.

A. nobilis, L. *Common Chamomile.* 6—9. Not common: woods and fields at Flixton, F.

CHRYSANTHEMUM, L.

C. segetum, L. *Corn Marigold.* 6—9. Rare: occasionally in fields between Dickleburgh and Billingford (DC).

C. Leucanthemum, L. *Great White Ox-eye.* 6—8. Abundant in pastures and on railway banks.

*C. Parthenium, Pers. (**Matricaria Parthenium**, L.). *Common Feverfew.* 6—9. Occasionally in hedge-banks: Baker's Barn, Redenhall; the Rectory Paddock (Miss Perowne), F. Flixton, I.

MATRICARIA, L.

M. inodora, L. *Scentless Mayweed.* 6—9. Plentiful in fields and waste places.

M. Chamomilla, L. *Wild Chamomile.* 6—8. Not common: roadside at Highgate, between Needham and Rushall, D. Waste ground, Shimpling, F.

TANACETUM, L.

T. vulgare, L. *Common Tansy.* 6—8. Abundant by roadsides and on banks of streams.

ARTEMISIA, L.

A. vulgaris, L. *Mug-wort.* 7—9. Common by waysides and in bushy places.

TUSSILAGO, L.

T. Farfara, L. *Common Coltsfoot.* 3—6. Common, especially in brickyards and on clay surfaces otherwise bare, C. *Cf.* Introd., p. 14, note.

PETASITES, Gaert.

*P. fragrans, Pres. *Sweet-scented Coltsfoot.* 1—3. Hedge-banks near Denton House, and between Pulham Market Station and Village, F. *Cf.* Introd., p. 26.

P. vulgaris, Desf. *Common Butter-bur.* 3—5. Frequent in wet places: beyond the island at Syleham; hedge-bank near Mendham Pit, on Withersdale Road; Mendham Mills; St. Margaret's, etc.

DORONICUM, L.

*D. Pardalianches, L. *Great Leopard's Bane.* 6, 7. Sparingly in the lane from Alburgh School to Denton, B

SENECIO, L.

S. vulgaris, L. *Common Groundsel.* 1—12. Generally distributed. Local name *Ascension Weed.*

S. sylvaticus, L. *Wood Groundsel.* 6—9. Not uncommon in bushy places: Gawdy Hall Wood; Shotford Heath, etc.

†S. viscosus, L. *Stinking Groundsel.* 7, 8. Very rare: in a gravel pit at Ditchingham (BG). *Cf.* Introd., p. 31.

S. erucifolius, L. *Hoary Ragwort.* 7—9. Frequent: Baker's Barn Brickyard, Redenhall: ditches, Wortwell Cricket Field: Tumbrill Hill, Needham; Bath Hills.

S. Jacobæa, L. *Common Ragwort.* 7—9. Common by roadsides and in waste places.

S. aquaticus, Huds. *Water Ragwort.* 7—9. Common in the marshes (Needham Alder Carr; Mendham, etc.). Dickleburgh Moor.

ARCTIUM, L.

A. majus, Schk. *Great Burdock.* 7—9. Frequent: Wood behind Weybread Hall; Well's Lane; Gawdy Hall Wood, etc.

A. minus, Schk. *Small Burdock.* 6—10. Common in fields and by roadsides.

A. intermedium, Lange. *Intermediate Burdock.* 7—9. Waste places, St. Margaret's (EAH).

CARDUUS, L.

C. nutans, L. *Musk Thistle.* 6—10. Frequent: Needham Alder Carr Pit; Wortwell; Shotford; Flixton; St. Margaret's, etc.

C. crispus, L. *Welted Thistle.* 6—9. Frequent: roadsides, Mendham, Wortwell, St. Margaret's, Dickleburgh; Harleston Green Lane. Gawdy Hall Wood (prob. var. **polyanthemos**).

CNICUS, Hoff.

C. lanceolatus, Hoff. *Spear Thistle.* 7—9. Common on banks and waste ground.

C. palustris, Hoff. *Marsh Thistle.* 7, 8. Common in wet meadows and damp woods. Frequently with white flowers.

C. pratensis, Willd. *Meadow Thistle.* 6—8. Very rare: damp places, Wacton Common (T).

C. acaulis, Hoff. *Dwarf Thistle.* 7—9. Rare: Balking Hill, Harleston, E. Bank near Redenhall Grange, B. Dickleburgh Moor (DC). Sandpit, St. Cross (EAH). Wacton Common (T).

C. arvensis, Hoff. *Creeping Plume Thistle.* 7, 8. Common by waysides and in cultivated ground. A variety with weaker spines by the side of Starston Rectory wall, F.

ONOPORDON, L.

O. Acanthium, L. *Scotch Thistle.* 7, 8. Frequent in meadows and waste places: field adjoining the White House, Harleston; Spurkett's Lane; near Mendham Mill. Alburgh, Brockdish (T). This Thistle, though adopted as the national emblem of Scotland, is quite a southern-type plant, and a doubtful native north of the Tweed.

SILYBUM, Gaert.

*S. Marianum, Gaert. (**Carduus Marianus**, L.). *Milk Thistle.* 7, 8. Frequent and well established: fir copse near Homersfield Church; between Mendham and Shotford Bridge (Norfolk); Brockdish. Alburgh (T).

CENTAUREA, L.

C. nigra, L. *Black Knap-weed.* 7—10. Common in fields and by waysides.

C. Scabiosa, L. *Greater Knap-weed.* 6—9. Not common: Needham Alder Carr Pit, and beside footpath from Gunshaws Hall to the Waveney, D. Mendham Gravel Pit, H.

CICHORIUM, L.

C. Intybus, L. *Wild Succory.* 6—9. Plentiful in cornfields and by waysides. The root dried and ground supplies the *Chicory* of commerce; the leaves are used as a salad under the name of *Endive.*

LAPSANA, L.

L. communis, L. *Common Nipplewort.* 7—9. Common in lanes and woods, and on the borders of meadows.

PICRIS, L.

P. hieracioides, L. *Hawkweed Ox-tongue.* 6—10. Frequent: by footpath and in pit on Shotford Hill; bushy places near Mendham Cross Roads; Norfolk; Dickleburgh; Flixton; Topcroft; St. Margaret's, etc.

P. echioides, L. (Helminthia echioides, Gaert.). *Bristly Ox-tongue.* 6—10. Not uncommon: Gawdy Hall Wood; near Mendham Grove, Norfolk; Denton Hill; St. Margaret's.

CREPIS, L.

C. virens, L. *Smooth Hawk's-beard.* 6—9. Common by roadsides and in waste places.

HIERACIUM, L.

H. Pilosella, L. *Mouse-ear Hawkweed.* 5—9. Common on dry banks and in pastures.

H. vulgatum, Fr. *Wood Hawkweed.* 7—9. Rare: copses, Dickleburgh (DC).

H. umbellatum, L. *Narrow-leaved Hawkweed.* 7—9. Not frequent: waste ground opposite Hulk's Graves, Weybread, E. Bath Hills (BG).

HYPOCHÆRIS, L.

H. glabra, L. *Smooth Cat's-ear.* 6—9. Rare: in a sandy field by the Bath Hills (NBG).

H. radicata, L. *Long-rooted Cat's-ear.* 6—8. Common on dry sandy and gravelly banks.

LEONTODON, L.

L. hirtus, L. (Thrincia hirta, DC.). *Hairy Hawk-bit.* 7—9. Not common: in footpath field adjoining Starston Rectory, B. Dickleburgh (DC). St. Margaret's (EAH), etc.

L. hispidus, L. (Apargia hispida, Willd.). *Rough Hawk-bit.* 6—9. Frequent in grassy places: Harleston Green Lane: Dickleburgh, etc.

L. autumnalis, L. (Apargia autumnalis, Willd.). *Autumnal Hawk-bit.* 8, 9. Frequent in pastures and waste ground: Harleston Green Lane; Mendham, etc.

TARAXACUM, Juss.

T. officinale, Web. *Common Dandelion.* 3—10. Var. densleonis, common. Var. erythrospermum, frequent on sandy soil: Beacon Hill, Shotford: Redenhall Road, near Lush Bush, F.

LACTUCA, L.

L. virosa, L. *Strong-scented Lettuce.* 7, 8. Rare: Shotford Heath Pit, C, D. Alburgh (T).

L. Scariola, L. *Prickly Lettuce.* 7, 8. Apparently frequent: Billingford, Needham, Harleston, Brockdish (T); Starston, Redenhall (TS). Have seen it only at Billingford, F. *Cf.* Introd., p. 28.

L. muralis, Fres. *Ivy-leaved Lettuce.* 6—8. Frequent in shady places: Starston; near Rushall Village: between Denton House and Rectory; Dickleburgh; Pulham Mary; St. Margaret's.

SONCHUS, L.

S. oleraceus, L. *Common Sow-Thistle.* 6—8. Common in fields and waste places.

S. asper, Hoff. *Rough Sow-Thistle.* 6—8. Common: Balking Hill, Harleston; Brockdish Road, etc.

S. arvensis, L. *Corn Sow-Thistle.* 8, 9. Plentiful in cornfields: Mendham Hill; Gawdy Hall; Flixton; Dickleburgh, etc.

TRAGOPOGON, L.

T. pratensis, L. *Yellow Goat's-beard.* 6, 7. Frequent: roadway, Clapper Farm: Mendham Hill: Ant Hill Farm, Redenhall: Gawdy Hall; Dickleburgh, etc. Var. **minor,** Redenhall Churchyard, F. Shimpling (T).

CAMPANULACEÆ.

JASIONE, L.

J. montana, L. *Annual Sheep's-bit.* 6—9. Not uncommon in sandy places: Mendham Pit, on Withersdale Road; roadside below Homersfield Church; opposite Hulk's Graves, Weybread; Needham Hill Pit.

CAMPANULA, L.

C. Trachelium, L. *Nettle-leaved Bell-flower.* 8—10. Rare: bushy places near Middleton Hall, Mendham, B. St. Margaret's (EAH).

C. latifolia, L. *Giant Bell-flower.* 7—9. Rare: thickets, Linstead Parva (BG). The Bath Hills (T).

C. rotundifolia, L. *Hare-bell.* 7—9. Frequent on dry banks: Harleston; Needham; Starston; Shotford Heath; waste ground opposite Hulk's Graves, Weybread; Homersfield Heath.

C. Rapunculus, L. *Rampion Bell-flower.* 7, 8. On old walls at Earsham (NBG).

C. patula, L. *Spreading Bell-flower.* 7—9. In hedges at Denton (Rev. C. T. Cruttwell), F.

SPECULARIA, DC.

S. hybrida, DC. *Venus' Looking-glass.* 6—9. Rare: in cultivated ground near Shotford Bridge, Norfolk, E.

ERICACEÆ.

CALLUNA, Salisb.

C. Erica, DC. (**C. vulgaris,** Sal.). *Common Ling.* 6—9. Very rare: one plant in the waste ground opposite Hulk's Graves, Weybread, E. *Cf.* Introd., p. 29.

MONOTROPEÆ.

HYPOPITYS, L.

H. **multiflora**, Scop. (**Monotropa Hypopitys**, L.). *Yellow Bird's-nest.* 7, 8. Rare: fir plantations in the Norfolk neighbourhood of Bungay (NBG). Beech woods at Earsham (Mr. F. Spalding).

PRIMULACEÆ.

HOTTONIA, L.

H. **palustris**, L. *Water Violet.* 6, 7. Abundant in the marsh dykes.

PRIMULA, L.

P. **vulgaris**, Huds. *Common Primrose.* 3—5. Common in woods and on hedge-banks. Var. **caulescens** (a luxuriant form with many flowers on one stalk) occasionally found.

P. **veris**, L. (**P. officinalis**, With.). *Cowslip.* 4, 5. Abundant in pastures.

P. **elatior**, Jacq. *Jacquin's Oxlip.* 4, 5. Rare: growing rather freely on a small boggy hill on Dickleburgh Moor (DC), F. *Cf.* Introd., p. 30.

P. **veri-vulgaris**, Syme. *Common Oxlip.* 4, 5. A hybrid. Frequently found with the Primrose and Cowslip.

LYSIMACHIA, L.

L. **vulgaris**, L. *Common Loosestrife.* 7, 8. Not common: near Spring Wood, Weybread, K. Shimpling (JM), K. Riverside between Shotford and Mendham (EAH). Formerly on Dickleburgh Moor (DC). At the foot of the Bath Hills (WA).

L. **Nummularia**, L. *Moneywort.* 6—8. Frequent in damp meadows: Shotford; Weybread; Mendham; Rushall; Dickleburgh; Pulham, etc. Popular name *Creeping Jenny.*

L. **nemorum**, L. *Yellow Pimpernel.* 6—9. Not uncommon in moist woods: Gawdy Hall Great Wood; Spring Wood, Weybread; Abbey Wood, Flixton; Billingford.

ANAGALLIS, Tour.

A. arvensis, L. *Scarlet Pimpernel.* 5—10. Common in cultivated ground. Popular name *Poor Man's Weather-glass.*

A. cærulea, Sch. *Blue Pimpernel.* 7, 8. Rare: cornfields at St. Margaret's (EAH), F. Redenhall (T).

A. tenella, L. *Bog Pimpernel.* 7, 8. Very rare: on Kett's Fen, Shimpling (JM).

SAMOLUS, Tour.

S. Valerandi, L. *Brookweed.* 7—9. Not common: riverside below Weybread Mill, C: rather plentiful near Luck's Mill, Needham, D. St. Margaret's (EAH). *Cf.* Introd., p. 32.

OLEACEÆ.

FRAXINUS, L.

F. excelsior, L. *Common Ash.* 4, 5. Common in woods and hedgerows.

LIGUSTRUM, Tour.

L. vulgare, L. *Common Privet.* 6, 7. Frequent in hedgerows near dwellings.

APOCYNACEÆ.

VINCA, L.

*V. major, L. *Greater Periwinkle.* 4—6. Hedge-bank near Weybread House, F. Pondside, Pulham Mary; Tivetshall: the Bath Hills (T).

V. minor, L. *Lesser Periwinkle.* 4—6. Abundant in some places: Mendham Grove, Norfolk; Denton; Flixton; St. Peter's; Earsham Wood, etc.

GENTIANEÆ.

BLACKSTONIA, Huds.

B. perfoliata, Huds. (Chlora perfoliata, L.). *Yellow Centaury.* 6—9. Not common: roadside between Flixton and St. Margaret's, I. Brookside, St. Margaret's (EAH). Hoxne (JC). Wacton and Earsham (T). Gravel pit adjoining the Bath Hills, with **Lathyrus Aphaca** (NBG).

ERYTHRÆA, Ren.

E. Centaurium, Pers. *Common Centaury.* 6—9. Frequent: Harleston Green Lane; Gawdy Hall Wood; lane from Rushall to Tumbrill Hill, Needham; Flixton, etc. At Hoxne, with pink, white, and red flowers (JC).

MENYANTHES, L.

M. trifoliata, L. *Common Buckbean.* 5—7. Not common: Stow Fen, near Earsham Mill (EAH), F. Shimpling (JM). Dickleburgh Moor, F. Formerly in the Wortwell Marshes (W. Squires).

BORAGINEÆ.

CYNOGLOSSUM, Tour.

C. officinale, L. *Hound's-tongue.* 6—8. Not infrequent: Mendham Grove and Pound Hole, Norfolk; Shotford Hill Pit; Earsham; St. Margaret's.

SYMPHYTUM, L.

S. officinale, L. *Common Comfrey.* 5—9. Not common: Mendham Pit, on Withersdale Road, E, H. Roadside between Weybread and Wingfield, F. Dickleburgh (DC).

BORAGO, L.

*B. officinalis, L. *Borage.* 6—8. Waste ground, Gawdy Hall Wood, I. Shimpling, on an old garden site (JM).

ANCHUSA, L.

*A. sempervirens, L. *Evergreen Alkanet.* 5—7. Rare: waste ground, Flixton Village, I.

LYCOPSIS, L.

L. arvensis, L. *Small Bugloss.* 6—8. Frequent on banks, by roadsides, and in fields: roadside below Homersfield Church, near Shotford Bridge, Norfolk. Weybread; Mendham; Flixton; Earsham; Needham; Billingford.

MYOSOTIS, Dill.

M. cæspitosa, Sch. *Tufted Water Forget-me not.* 6—8. Not common: ditches of the Waveney at Mendham, F.

M. palustris, With. *Common Water Forget-me-not.* 6—9. Abundant in the Waveney and in moist places.

M. repens, Don. *Creeping Water Forget-me-not.* 6—9. Not common : St. Margaret's (EAH).

M. sylvatica, Hoff. *Wood Forget-me-not.* 5—8. Rare : Flixton Woods (EAH), I.

M. arvensis, Hoff. *Field Forget-me-not.* 6—8. Common in fields, woods, and waste places.

M. collina, Hoff. *Dwarf Forget-me-not.* 4, 5. Common on dry banks in the spring.

M. versicolor, Reich. *Yellow and blue Forget-me-not.* 4—6. Not common : wayside bank, Homersfield Wood (EAH); Homersfield Heath and Flixton, F.

LITHOSPERMUM, Tour.

L. officinale, L. *Common Gromwell.* 5—8. Frequent in waste places : Redenhall Green Lane ; Gawdy Hall Wood ; Billingford ; St. Margaret's, etc.

L. arvense, L. *Corn Gromwell.* 5—7. Frequent in cultivated fields: near Starston Railway Bridge ; near Spurkett's Lane ; near Mendham Old Priory ; St. Margaret's, etc.

ECHIUM, L.

E. vulgare, L. *Viper's Bugloss.* 6—8. Not uncommon in gravelly places : Needham Alder Carr Pit : near Harleston Bridge ; Flixton New Road ; Dickleburgh.

CONVOLVULACEÆ.

CALYSTEGIA, R. Br.

C. Sepium, R.Br. **(Convolvulus sepium, L.).** *White Convolvulus.* 7, 8. Common in bushy places, especially near water-courses. Popular name *Bindweed*.

CONVOLVULUS, L.

C. arvensis, L. *Field Convolvulus.* 6—8. Common on banks and in fields.

CUSCUTA, Tour.

†C. **Epilinum**, Weihe. *Flax Dodder.* 7. Shimpling (JM). This parasitical plant is probably extinct, as Flax is no longer grown in the district.

C. **Epithymum**, Murr. *Lesser Dodder.* 7, 8. On furze and thyme: rare: Dickleburgh (DC). Also near Bungay (T).

*C. **Trifolii**, Bab. *Clover Dodder.* 7—9. Occasional: cloverfield near Old Weybread Green, D. Field near the New Buildings, Harleston, K. St. Margaret's (EAH).

SOLANACEÆ.

SOLANUM, L.

S. **Dulcamara**, L. *Woody Nightshade.* 6—8. Frequent in hedgerows, and common in the marshes. Popular name *Bittersweet*. Berries poisonous.

S. **nigrum**, L. *Black Nightshade.* 6—10. Abundant in waste places: Clapper Pit, Mendham; marsh above Syleham Mill; Harleston Gasworks; Brockdish; Dickleburgh, etc. Berries poisonous.

LYCIUM, L.

*L. **barbarum**, L. *Box-thorn.* 6—9. In hedgerows near dwellings: Redenhall; Dickleburgh, etc.

ATROPA, L.

*A. **Belladonna**, L. *Deadly Nightshade.* 6—8. A weed in the late Mr. Muskett's garden and yard, Harleston, F. Plentiful at Framlingham Castle. The most poisonous of all British plants to human beings, though animals and birds appear unaffected by it.

HYOSCYAMUS, L.

H. **niger**, L. *Common Henbane.* 5—8. Occasional: railway cutting near Lush Bush (the late Mr. Muskett). A weed in garden ground, Dickleburgh (DC). Earsham (T); St. Margaret's (EAH). Mr. Muskett observed that the seeds require repeated exposure to the air in order to insure growth, and will lie dormant for years until the soil is newly turned. Poisonous.

SCROPHULARINEÆ.

VERBASCUM, L.

V. Thapsus, L. *Great Mullein.* 6—9. Common on dry banks: roadside, Shotford to Mendham, Norfolk; Shotford Heath; Shimpling; Billingford, etc.

V. pulverulentum, Vill. *Hoary Mullein.* 7. Rare: roadside at Earsham (Mr. F. Spalding). Ditchingham (WA).

V. nigrum, L. *Dark Mullein.* 6—9. Frequent by waysides: Homersfield and St. Cross Road; Mendham and Metfield Road; Scole; Billingford; Great Melton, etc.

V. Blattaria, L. *Moth Mullein.* 7, 8. Rare: waste ground, Weybread (JH, 1863).

LINARIA, Tour.

*L. **Cymbalaria,** Mill. *Ivy-leaved Toad-flax.* 6—10. Frequent: walls; Wilderness Lane, Harleston; Brockdish; Flixton, etc.

L. Elatine, Mill. *Sharp-leaved Fluellin.* 7, 8. Not common: field, Harleston Green Lane, opposite Anthill Farm, F. Field near North Lodge, Weybread, D. Dickleburgh (DC), F. St. Margaret's (EAH).

L. spuria, Mill. *Round-leaved Fluellin.* 7—9. Not common: with **L. Elatine,** near the Green Lane, Harleston, F. Field near Mendham Grove, Norfolk. At St. Cross in great abundance (EAH).

L. vulgaris, Mill. *Yellow Toad-flax.* 7—9. Frequent on dry banks and in cornfields.

L. viscida, Mœnch. (**L. minor,** Desf.). *Least Toad-flax.* 6—10. Not common: cornfields, St. Margaret's (EAH), F. Flixton I.

ANTIRRHINUM, L.

A. Orontium, L. *Lesser Snap Dragon.* 6—10. Not common: Homersfield Allotments (W. Squires), F. Wortwell School Allotments, F. Cultivated ground, Dickleburgh (DC).

SCROPHULARIA, L.

S. aquatica (var. **Balbisii**), L. *Water Betony.* 6—9. Abundant in the marsh dykes and by the side of streams.

S. nodosa, L. *Knotted Figwort.* 6—8. Frequent in damp bushy places: Gawdy Hall Wood; Alburgh; marshes at Weybread and Shotford, etc.

VERONICA, Tour.

V. hederifolia, L. *Ivy-leaved Speedwell.* 3—8. Common in fields and by waysides. Popular name *Winterweed.*

V. polita, Fr. *Grey Procumbent Speedwell.* 3—9. Frequent in fields and waste places: Harleston Green Lane; Redenhall Road; London Road, Harleston; St. Margaret's; Dickleburgh, etc.

V. agrestis, L. *Green Procumbent Speedwell.* 3—9. Common in fields and on banks: Harleston Bridge; Flixton, etc.

*V. persica, Poir. (V. Buxbaumii, Ten.). *Buxbaum's Speedwell.* 4—9. Abundant in cultivated fields. This strong species, introduced into England in 1829, is rapidly exterminating V. polita and V. agrestis.

†V. verna, L. *Vernal Speedwell.* 4, 5. Balking Hill, Harleston (BG). The features of this locality are now altered through cultivation and enclosure.

V. arvensis, L. *Wall Speedwell.* 4—7. Common in pastures and waste places. A very small form on Balking Hill, F.

V. serpyllifolia, L. *Thyme-leaved Speedwell.* 5—7. Frequent in dry places and by roadsides: Shotford Heath; Spurkett's Lane; Flixton; Homersfield Wood; Dickleburgh.

V. officinalis, L. *Common Speedwell.* 5—8. Frequent: Gawdy Hall Great Wood; near Weybread Mill: Mendham; Needham; Flixton; Brockdish; Dickleburgh, etc.

V. Chamædrys, L. *Germander Speedwell.* 5—8. Common on grassy banks and in woods. Popular name *Bird's-eye.*

V. montana, L. *Mountain Speedwell.* 4—6. Rare: roadside near Flixton Church and at St. Cross (EAH). Flixton Woods, I. Earsham (T).

V. scutellata, L. *Marsh Speedwell.* 6—8. Not common: marshy ground, Wingfield Common, D. Boggy Meadow, St. Margaret's (EAH), F.

V. Anagallis, L. *Water Speedwell.* 6—8. Frequent in marsh ditches: Mendham; Weybread; Flixton; Scole; Dickleburgh, etc.

V. Beccabunga, L, *Brooklime.* 5—9. Common in ponds and ditches.

EUPHRASIA, L.

E. officinalis, L. *Common Eyebright.* 5—10. Frequent: Clintergate Road, Redenhall; Mendham Hill, Norfolk; Stubbings' Lane, Weybread; Flixton Park; Dickleburgh, etc.

BARTSIA, L.

B. Odontites, Huds. *Red Bartsia.* 7, 8. Frequent in cultivated ground: Gawdy Hall; Dickleburgh, etc.

PEDICULARIS, L.

P. sylvatica, L. *Procumbent Lousewort.* 5—8. Rare: between Brockdish and Rushall, near Fir Cover (EAH).

MELAMPYRUM, L.

M. pratense, L. *Cow-wheat.* 5—8. Rare: Gawdy Hall Great Wood, F. Knight's Grove, Langmere (DC). Billingford (JC).

RHINANTHUS, L.

R. Crista-galli, L. *Yellow Rattle.* 5—7. Common in damp meadows: Weybread; Shotford; Wortwell; Flixton; Dickleburgh, etc.

OROBANCHACEÆ.

OROBANCHE, L.

†O. ramosa, L. *Branched Broom-rape.* 8, 9. Earsham (T). Formerly not uncommon on Hemp and *Galeopsis Tetrahit.* Mr. Holmes, however, failed to find it on Hemp grown some years ago by the late Lord Waveney at Flixton.

O. major, L. (O. Rapum, Thuill.). *Greater Broom-rape.* 6, 7. Not common: waste ground opposite Hulk's Graves, Weybread, D. The Bath Hills (T). Dickleburgh (DC). Parasitical on broom and furze.

O. elatior, Sut. *Tall Broom-rape.* 6—8. Rare: Needham (T). Parasitical on Knapweed.

O. minor, Sm. *Lesser Broom-rape.* 6—8. Frequent: fields near Harleston Bridge; Wortwell; Pulham; Shimpling; Flixton, etc. Parasitical on Clover and many other plants, including the garden Geranium.

LENTIBULARIEÆ.

UTRICULARIA, L.

U. vulgaris, L. *Greater Bladderwort.* 6—8. Now rare: pond between Brockdish and Rushall, B. Ditch adjoining Needham Osier Ground (the late Mr. Muskett). In ditches near Mendham Bridge and at Brockdish (EAH). *Cf.* Introd., p. 29.

U. minor, L. *Lesser Bladderwort.* 6—8. Very rare: in slow water at Pulham Market (TS).

VERBENACEÆ.

VERBENA, L.

V. officinalis, L. *Common Vervain.* 7—9. Frequent by roadsides: Redenhall; Well's Lane, Harleston; Wortwell Low Street; Flixton; Dickleburgh, etc.

LABIATÆ.

MENTHA, L.

M. rotundifolia, L. *Round-leaved Mint.* 7—9. Not infrequent in banks and moist places: roadside opposite Mendham Priory Mansion; Withersdale Street; Metfield Road; near Weybread Windmill; Shimpling Common; St. Cross.

M. sylvestris, L. *Horse Mint.* 7—9. Rare: road from Metfield Parsonage to Church, at the bottom of the hill (EAH). Moist places, Pulham Market (TS).

*M. viridis, L. *Spear Mint.* 8—10. Rare: on the south side of the Waveney, about a mile below Syleham Mill (EAH). Alburgh (TS).

M. Piperita, Huds. *Peppermint.* 8, 9. Rare: damp places: var. officinalis, Denton (TS). Var. vulgaris, Harleston (TS).

M. hirsuta, L. *Hairy Capitate Mint.* 7—9. Common in ditches and damp places. Often very luxuriant, as at Mendham Cross Roads, Norfolk.

M. sativa, L. *Hairy Whorled Mint.* 7—9. Not common: Pond, Harleston Common; ditches, Weybread Marshes; Gawdy Hall Wood, F. Scole, Shimpling (TS).

M. rubra, Sm. *Glabrous Red Mint.* 7—9. Rare: damp situations, Wacton and Scole (TS). Reported from the Waveney Marshes, but requires confirmation.

M. gentilis, L. *Bushy Red Mint.* 7—9. Rare: damp places, Pulham, Starston, Shelton, Hempnall (TS).

M. arvensis, L. *Corn Mint.* 8, 9. Common in cultivated ground. Var. **Allionii**, Needham (TS).

M. Pulegium, L. *Penny-royal.* 8, 9. Rare: wet places, Bedingham (T). Ditchingham (BG).

LYCOPUS, L.

L. europæus, L. *Water Horehound.* 6—9. Frequent in wet places: Harleston Green Lane; Gawdy Hall Wood; Dickleburgh Moor. Abundant in the marsh dykes. Popular name *Gipsywort*.

ORIGANUM, L.

O. vulgare, L. *Sweet Marjoram.* 7—9. Rare: dry banks, Hempnall (TS). Var. **megastachyum**, Hardwick (TS). This is supposed to be the *Hyssop* of the Bible.

THYMUS, L.

T. Serpyllum, Fr. *Creeping Wild Thyme.* 6—8. Not common: Flixton New Road, I.

CALAMINTHA, Mœnch.

C. Clinopodium, Benth. *Wild Basil.* 7—9. Frequent in bushy places: Harleston Green Lane; Lush Bush; Mendham Priory Plantations; Starston, etc.

C. officinalis, Mœnch. (**C. menthifolia**, Host.). *Common Calamint.* 7—9. Frequent on dry banks: Shotford Hill; Withersdale Road, Mendham; Metfield; Homersfield, etc.

SALVIA, L.

S. Verbenaca, L. *Wild English Clary.* 5—8. Abundant on dry banks: Wortwell; Well's Lane, Harleston; Starston; Billingford; Scole; Shotford; Flixton; St. Margaret's; Ditchingham, etc.

NEPETA, L.

N. Cataria, L. *Cat-mint.* 7—9. Frequent on banks: Wortwell End; roadside near Shotford Hall; Abbey Wood, Flixton; Dickleburgh, etc.

N. Glechoma, Benth. *Ground-ivy.* 3—6. Common in hedges and woods. Before the introduction of Hops the Groundivy, with the Wood Sage and Sweet Marjoram, was in great demand for ale-brewing; hence its popular name *Ale-hoof.*

SCUTELLARIA, L.

S. galericulata, L. *Sku'l-cap.* 7, 8. Abundant in the Waveney Marshes. Gawdy Hall Wood; Shimpling.

PRUNELLA, L.

P. vulgaris, L. *Self-heal.* 7—9. Common in meadows and waste places.

MARRUBIUM, L.

M. vulgare, L. *White Horehound.* 7—9. Rare; waste ground, Brockdish (T). On the site of St. Nicholas Church, South Elmham (extinct, EAH).

STACHYS, L.

S. Betonica, Benth. *Wood Betony.* 7, 8. Rare: pastures, Brockdish (EAH). Footpath field near Syleham Hall (W. Squires).

S. palustris, L. *Marsh Woundwort.* 7—9. Common in the water meadows: also in cornfields and by roadsides.

S. sylvatica, L. *Hedge Woundwort.* 7—9. Common in hedgebanks and bushy places.

S. arvensis, L. *Corn Woundwort.* 4—10. Frequent in cultivated fields: Mendham Hill, Norfolk; fields near Shotford Bridge; Lush Bush Allotments; near the Woodman's Cottage, Gawdy Hall; Brockdish, etc.

GALEOPSIS, L.

G. Ladanum, L. *Red Hemp-nettle.* 6—9. Not common: in cultivated ground near the river, Brockdish, F.

G. speciosa, Mill. (G. versicolor, Curt.). *Large-flowered Hemp-nettle.* 7, 8. Rare: cultivated fields, Dickleburgh (DC). Earsham (T). Brockdish (not lately, EAH). Eye (JH, 1876).

G. Tetrahit, L. *Common Hemp-nettle.* 7—9. Common in cornfields and cultivated ground, especially in damp situations. With white flowers in cornfields below Mendham Mill and near the Shotford Dykes.

LEONURUS, L.

*****L. Cardiaca**, L. *Motherwort.* 7, 8. Rare: hedges and waste places: near (the late) Wortwell Windmill (BG), but probably extinct, F. Denton, Ditchingham (T).

LAMIUM, L.

L. amplexicaule, L. *Henbit Dead-nettle.* 5—9. Not common: gravel pit near Redenhall Gatehouse, F.

L. hybridum, Vill. (L. incisum, Willd.). *Cut-leaved Dead-nettle.* 4—8. Frequent: roadside near Weybread House; the Hol-Way, Gawdy Hall, Needham, etc.

L. purpureum, L. *Red Dead-nettle.* 5—9. Common in hedges and waste places. With white flowers near Shotford Hall and Mendham Mill. Var. **decipiens**, waste ground, Withersdale, F.

L. album, L. *White Dead-nettle.* 5—8. Abundant in hedgebanks.

L. Galeobdolon, Crantz. *Yellow Archangel.* 5, 6. Frequent in bushy places: Cuckoo Hill, Mendham, Norfolk. Lush Bush; Homersfield Wood; Billingford; Dickleburgh; Bath Hills, etc.

BALLOTA, L.

B. nigra, L. *Black Horehound.* 7—9. Common in hedgebanks.

TEUCRIUM, L.

T. scorodonia, L. *Wood Sage.* 7, 8. Not common: bushy places opposite Hulk's Graves, Weybread, D. Mendham Priory Plantations, F. Foxburrows Plantation, Weybread, C, D.

AJUGA, L.

A. reptans, L. *Common Bugle.* 5—8. Common in pastures and woods.

PLANTAGINEÆ.

PLANTAGO, L.

P. major, L. *Greater Plantain.* 6—9. Common by roadsides and in waste places.

P. media, L. *Hoary Plantain.* 6—9. Common in pastures.

P. lanceolata, L. *Rib-wort Plantain.* 6—9. Common in waste places.

P. Coronopus, L. *Buck's-horn Plantain.* 6—8. Frequent in waste places: Wortwell Road, opposite the Schoolroom; Well's Lane, Harleston; Needham Alder Carr, etc.

ILLECEBRACEÆ.

SCLERANTHUS, L.

S. annuus, L. *Annual Knawel.* 6—8. Rare: a few plants on sandy ground, opposite Hulk's Graves, Weybread, F. Abundant on Bungay Common.

CHENOPODIACEÆ.

CHENOPODIUM, L.

C. polyspermum, L. *Many-seeded Goosefoot.* 7—9. Abundant in cultivated ground.

C. album, L. *White Goosefoot.* 7—9. Var. **candicans**, common. Var. **viride**, Wortwell School Allotments, F. Var. **paganum**, Allotments near Dickleburgh Rectory, F.

C. murale, L. *Nettle-leaved Goosefoot.* 6—9. Rare: waste ground, St. Margaret's (EAH).

C. hybridum, L. *Maple-leaved Goosefoot.* 7. Rare: waste places: St. Margaret's (EAH). Earsham (T).

C. glaucum, L. *Oak-leaved Goosefoot.* 7, 8. Rare: waste ground, Ditchingham (TS).

C. Bonus-Henricus, L. *Allgood.* 6—8. Abundant by roadsides and in waste places, especially near villages: Redenhall; Wortwell End; Dickleburgh, etc. Popular name *Good King Harry.*

G

ATRIPLEX, L.

A. patula, L. (Var. **angustifolia,** Sm.). *Spreading narrow-leaved Orache.* 7—10. Frequent on cultivated and waste ground: Needham Alder Carr Pit: St. Margaret's; roadside, Redenhall, etc.

Var. **erecta,** rare: Brockdish (EAH).

A. hastata, L. (**A. Smithii,** Sm.). *Hastate-leaved Orache.* 7—9. Not uncommon: waste ground below Balking Hill, Harleston; Wortwell; fields between Weybread and Needham Mills, etc.

A. deltoidea, Bab. *Triangular-leaved Orache.* 6—8. Frequent in waste places and cultivated land: Harleston; Homersfield Allotments, etc.

POLYGONACEÆ.

POLYGONUM, L.

P. Convolvulus, L. *Climbing Buck-wheat.* 7—9. Abundant on cultivated land.

P. aviculare, L. *Common Knotgrass.* 5—9. Common in waste places.

P. Hydropiper, L. *Water Pepper.* 7—9. Common in ditches and marshes.

P. minus, Huds. *Small Persicaria.* 8, 9. Rare: damp places in the neighbourhood of St. Margaret's (EAH).

P. Persicaria, L. *Spotted Persicaria.* 6—9. Common in damp places and in cultivated ground.

P. lapathifolium, L. *Glandular Persicaria.* 7, 8. Not infrequent: damp ground near Weybread Targets; beckside near Redenhall Gatehouse, F. Needham (T). St. Margaret's (EAH).

P. amphibium, L. *Amphibious Bistort.* 7—9. Common in or near the marsh dykes.

P. Bistorta, L. *Common Bistort.* 6, 7. Not common: on the island, Gawdy Hall Great Wood (Mrs. Pemberton), F. in meadows near the Church, Mendham; Flixton Hollow (EAH).

RUMEX, L.

R. conglomeratus, Murr. *Sharp Dock.* 6—9. Frequent in the marshes.

R. sanguineus, L. (R. nemorosus, Sch.). *Bloody-veined Dock.* 6, 7. Var. viridis, not unfrequent: Needham Alder Carr; Gawdy Hall Wood; Flixton Park.

R. pulcher, L. *Fiddle Dock.* 7—9. Frequent by roadsides: Redenhall; Well's Lane, Harleston; Flixton, etc.

R. obtusifolius, L. *Broad-leaved Dock.* 5—10. Common by waysides and in fields.

R. crispus, L. *Curled Dock.* 6—9. Frequent in waste ground: London Road, Harleston; The Green Lane, etc.

R. Hydrolapathum, Huds. *Great Water Dock.* 7, 8. Abundant by the riverside and in the dykes.

R. Acetosa, L. *Common Sorrel.* 5—7. Common in meadows and woods.

R. Acetosella, L. *Sheep's Sorrel.* 5—7. Common on dry banks and in pastures.

THYMELÆACEÆ.

DAPHNE, L.

*D. Mezereum, L. *Mezereon.* 3. The Lady's Grove, Gawdy Hall (EAH). Formerly at Laxfield (WA). Ditchingham (T).

D. Laureola, L. *Spurge Laurel.* 2—5. Frequent in woods and hedge-banks: Baker's Barn, Redenhall; Mendham Grove, Norfolk; near St. Cross Schoolroom; Flixton; the Bath Hills; Denton Plantations; Redenhall Green Lane; Shelton; Dickleburgh; Hoxne, etc.

LORANTHACEÆ.

VISCUM, L.

V. album, M. *Mistletoe.* 3—5. Not common: on poplar-trees, Pulham Market; on Wild Apple, Pulham Mary, F. Orchards, Homersfield, G. Alburgh, F. St. Margaret's (EAH). Veales Farm, Fressingfield, N.

EUPHORBIACEÆ.
EUPHORBIA, L.

E. Helioscopia, L. *Sun Spurge.* 3—10. Common in cultivated ground.

E. amygdaloides, L. *Wood Spurge.* 4—8. Rare: bushy places, Shelton (TS).

E. Peplus, L. *Petty Spurge.* 6—9. Common in cultivated ground.

E. exigua, L. *Dwarf Spurge.* 7—10. Common in cultivated fields. Var. **retusa,** DC. Shimpling, F.

*E. Lathyris, L. *Caper Spurge.* 6, 7. In waste places, and a weed in old gardens, Harleston, Withersdale, F.

MERCURIALIS, L.

M. perennis, L. *Perennial Dog's Mercury.* Common in woods and shady banks.

URTICACEÆ.
ULMUS, L.

U. montana, Sm. *Broad-leaved Elm.* 3, 4. Redenhall Road, near the second railway bridge, F. Popular name *Wych Elm.*

U. campestris, Sm. *Common Elm.* 3—5. Var. **suberosa,** common.

HUMULUS, L.

H. Lupulus, L. *Common Hop.* 7—9. Frequent in hedgerows: Harleston; Redenhall; Denton; Flixton; Withersdale; Mendham, etc.

URTICA, L.

U. dioica, L. *Common Nettle.* 5—9. Common in waste places and hedge-banks.

U. urens, L. *Small Nettle.* 5—9. Abundant in waste places and cultivated ground.

PARIETARIA, Tour.

P. officinalis, L. *Pellitory of the Wall.* 5—10. Not common: walls of Wingfield Castle, F. Metfield churchyard wall, D. Pulham (T).

CUPULIFERÆ.

BETULA, Tour.

B. alba. L. *White Birch.* 4, 5. Not abundant. A row of fine trees on the Weybread and Syleham Road, C. "*Byrche* is called in Latin **Betula**. I have not red of any vertue that it hath in physik, howbeit it serveth for many good uses, and none better than in betinge stubborne boys that ether lye or wyll not learne."—TURNER, 1550. (Quoted in Flor. Dorset.)

ALNUS, Tour.

A. glutinosa, L. *Common Alder.* 4, 5. Abundant in damp thickets, and by banks of streams.

CARPINUS, L.

C. Betulus, L. *Hornbeam.* 4, 5. Abundant in hedgerows and woods: Gawdy Hall Great Wood; Harleston Green Lane; St. Margaret's, etc.

CORYLUS, Tour.

C. Avellana, L. *Hazel.* 2—4. Common in hedgerows and woods.

QUERCUS, L.

Q. Robur, L. *Oak.* 4, 5. Var. **pedunculata**, common. A very fine tree near the railway (north side), between Pulham Market and Tivetshall Stations.

CASTANEA, Tour.

*C. sativa, Mill. (C. **vulgaris**, Lam.). *Sweet Chestnut.* 5, 6. Priory Plantations, Mendham. Fressingfield Hall Plantations.

FAGUS, Tour.

F. sylvatica, L. *Common Beech.* 5. Abundant in woods and hedgerows.

SALICINEÆ.

SALIX, Tour.

S. fragilis, L. *Crack Willow.* 4, 5. Frequent in moist meadows: near Weybread Mill, etc.

S. alba, L. *White Willow.* 4. Frequent on banks of streams: near Needham Osier Ground; Starston Beck; Wortwell, etc. A fine male tree near Spring Wood Bridge, Weybread.

S. triandra, L. *Almond-leaved Willow.* 4—6. Abundant on damp banks. Cultivated in osier grounds.

S. purpurea, L. *Purple Willow.* 3—5. Not common: riverside near Weybread Mill; Shimpling, F.

S. rubra, var. **Forbyana**, Sm. *Basket Osier.* 4. Rare: riverside at Needham Osier Ground. Many varieties are cultivated in the district for basket-work.

S. viminalis, L. *Common Osier.* 4, 5. Common in damp places and osier grounds.

S. cinerea, L. *Common Sallow.* 3—5. Not common: Shotford Heath Pit, F. Var. **aquatica**, common in moist places: Mendham Long Lane; Gawdy Hall Wood, etc. Var. **oleifolia**, frequent: Harleston Green Lane, etc.

S. Caprea, L. *Great Sallow.* 4, 5. Frequent in hedges and woods: Harleston Green Lane; Wortwell; Gawdy Hall Wood, etc.

POPULUS, Tour.

P. alba, L. *White Poplar.* 3. Not common: Mendham Grove, Norfolk; Pulham Market, F. St. Margaret's (EAH). Popular name *Abele.*

P. canescens, Sm. *Grey Poplar.* 3, 4. Frequent: roadside near Redenhall Church; Shotford; Starston, etc.

P. tremula, L. *Aspen.* 4, 5. Frequent: Shotford Hill; Wortwell; Gawdy Hall Wood.

*P. nigra, L. *Black Poplar.* 4. Abundant: Shotford Bridge; Homersfield; Redenhall; Pulham; Rushall, etc.

CERATOPHYLLEÆ.

CERATOPHYLLUM, L.

C. demersum, L. *Common Hornwort.* 6—8. Frequent in the marsh dykes: Homersfield; Shotford; Mendham, etc. Ditches, Lush Bush.

C. submersum, L. *Unarmed Hornwort.* 6—8. Rare: ditches, Brockdish (EAH).

CONIFERÆ.

TAXUS, Tour.

*__T. baccata,__ L. *Yew.* 3. Occasionally in plantations. An old tree opposite the Yew Tree Inn, Redenhall.

PINUS, Tour.

*__P. sylvestris,__ L. *Scotch Fir.* 5, 6. On sandy and gravelly soil: Homersfield; Redenhall; Shotford; Earsham, etc.

MONOCOTYLEDONES.

HYDROCHARIDEÆ.

ELODEA, Mich.

*E. canadensis, Mich. (Anacharis Alsinastrum, Bab.). *Water Thyme.* 7—9. Frequent in the marsh dykes: Wortwell, Shotford, etc. A North American plant, first observed in England in 1847, and now generally distributed.

HYDROCHARIS, L.

H. Morsus-ranæ, L. *Frog-bit.* 7, 8. Frequent in ditches: Weybread Mill; Shotford Bridge; Mendham Marshes; Dickleburgh, etc.

STRATIOTES, L.

S. aloides, L. *Water-Soldier.* 7, 8. Formerly frequent, now rare: Homersfield ditches (1863), and slow ditch above Syleham Mill, Norfolk (EAH). Pond in Alder Carr Meadow, Needham (the late J. Muskett). Ditches, Scole and Billingford (JC). *Cf.* Introd., p. 29.

ORCHIDEÆ.

NEOTTIA, L.

N. Nidus-avis, Rich. *Bird's-nest Orchis.* 6. Not common: Gawdy Hall Wood (BG), I. Barker's Wood, Rushall (EAH). Hedenham and Tindall Woods (T).

LISTERA, R. Br.

L. ovata, R. Br. *Common Tway Blade.* 5—7. Abundant in woods and copses: Gawdy Hall; Denton; Starston; Weybread; Needham; the Bath Hills, etc.

SPIRANTHES, Rich.

S. autumnalis, Rich. *Autumnal Lady's Tresses.* 7—9. Rare: St. Margaret's and All Saints (EAH). Flixton (the late J. Muskett). Near Gawdy Hall Wood (Mrs. Pemberton). Bedingham Green (BG).

CEPHALANTHERA, Rich.

†C. ensifolia, Rich. *Long-leaved Helleborine.* 5, 6. Rare: bushy places at the foot of the Bath Hills (BG).

EPIPACTIS, Rich.

E. latifolia, All. *Broad-leaved Helleborine.* 7, 8. Rare: the Abbey Wood, Flixton, E. Gawdy Hall Wood (Mrs. Pemberton). Reported also from Starston.

ORCHIS, L.

O. pyramidalis, L. *Pyramidal Orchis.* 6—8. Frequent: Redenhall; Starston; Baker's Barn; Gawdy Hall; Alburgh; Flixton; Mendham; Hoxne; Scole; Dickleburgh; Shimpling; Gissing; Tivetshall; Shelton, etc.

O. Morio, L. *Green-winged Orchis.* 5, 6. Abundant in meadows: occasionally with white flowers. This and the next species are called *Cuckoo-flower* in Norfolk.

O. mascula, L. *Early Purple Orchis.* 4, 5. Common in woods and meadows.

O. incarnata, L. *Common Marsh Orchis.* 5, 6. Not common: marshy field, St. Margaret's (EAH), and meadows near Weybread Mill, E.

O. latifolia, L. *Broad-leaved Marsh Orchis.* 5, 6. Frequent: Wortwell Marshes; Weybread Marshes near the Targets; St. Margaret's; Billingford; Wacton; Dickleburgh.

O. maculata, L. *Spotted Palmate Orchis.* 5—7. Common in woods and damp places.

ACERAS, R. Br.

A. anthropophora, R. Br. *Green Man-Orchis.* 6. Not common: meadows near Mendham Long Lane (the late J. Muskett). Hartcup's plantations near Bungay (EAH).

Shimpling (JM). Tivetshall St. Margaret (TS). Earsham (NBG). In a dry pit (now the Dove House Dell) at the end of Mr. Wright's garden at Mendham, Norfolk (BG).

OPHRYS, L.

O. apifera, Huds. *Bee Orchis.* 6, 7. Common in some years : Needham Alder Carr Meadows; between Weybread Church and Beck; Baker's Barn Brickyard, Redenhall : Gawdy Hall Wood; Mendham Long Lane; Denton; Earsham ; St. Margaret's ; Flixton ; Dickleburgh ; Shimpling. Frequent near Harleston (BG, 1805). Plentiful 1884.

O. muscifera, Huds. *Fly Orchis.* 5—7. Occasional : All Saints' Rectory field with **Spiranthes autumnalis** (EAH, 1884). Shimpling (JC). Gawdy Hall Wood (NBG, and Mrs. Pemberton). Bath Hills and Earsham Wood (WA).

HABENARIA, R. Br.

H. conopsea, Benth. (**Gymnadenia conopsea,** R. Br.). *Fragrant Orchis.* 6—8. Not common : Shimpling (JM). Dickleburgh (DC). Pulham (TS). Sparingly on Wacton Common (T). Reported also from Gawdy Hall.

H. viridis, R. Br. *Frog Orchis.* 6—8. Abundant in fields near St. Margaret's (EAH), F. Ditchingham (T).

H. bifolia, R. Br. *Lesser Butterfly Orchis.* 6—8. Frequent in damp woods and pastures : Gawdy Hall Wood; Needham ; Weybread ; Starston ; Mendham. Denton (Rev. C. T. Cruttwell). Wacton (T).

H. chloroleuca, Rid. (**H. chlorantha,** Bab.). *Greater Butterfly Orchis.* 5, 6. Not uncommon : Gawdy Hall Wood; Starston Wood ; Rushall Wood ; Flixton.

IRIDEÆ.

IRIS, L.

I. fœtidissima, L. *Blue Iris.* 5—7. Not common : Mendham Grove, Norfolk, F. Flixton Long Plantation, I. Swampy ground, Tivetshall Wood (Rev. H. T. Frere). Very abundant on the Bath Hills, Ditchingham (WA). *Cf.* Introd., p. 32.

I. **Pseudacorus,** L. *Yellow Iris.* 5—8. Frequent: banks of the Waveney; Redenhall Beck; pond, Shotford Hill; Spring Wood, Weybread; Flixton, etc. Popular name *Flag.*

CROCUS, Tour.

* **C. vernus,** All. *Spring Crocus.* 3. It "covers a field by the side of Mendham Long Lane by Harleston, and has grown there before the memory of the oldest person in the neighbourhood" (BG, 1805). Still abundant and apparently spreading.

AMARYLLIDEÆ.

NARCISSUS, L.

N. **Pseudo-narcissus,** L. *Wild Daffodil.* 3, 4. Not common: occasionally in meadows near Jay's Green, Harleston, F. Plentiful on the island, Gawdy Hall Wood (EAH), F. Popular name *Lent Lily.*

*N. **major,** L. *Spanish Daffodil.* 3, 4. Mendham Grove, Norfolk, in the last century a garden. *Cf.* Introd., p. 29, note.

*N. **biflorus,** Curt. *Two-flowered Narcissus.* 4, 5. Rare: in a meadow at St. Margaret's (EAH), E. Mendham Grove, Norfolk, DH. *Cf.* Introd., p. 26.

GALANTHUS, L.

G. **nivalis,** L. *Common Snowdrop.* 2, 3. Not common: marshy corner of a meadow by the Waveney, Weybread, C. Plentiful in the Long Plantation, Flixton (EAH), F. Hedges at Laxfield in the greatest profusion (BG).

DIOSCOREÆ.

TAMUS, L.

T. **communis,** L. *Black Bryony.* 5—7. Frequent in woods and hedges: Jay's Green, Harleston; Gawdy Hall Wood; Flixton; Dickleburgh; Thorpe Abbotts, etc.

LILIACEÆ.

RUSCUS, L.

R. **aculeatus,** L. *Butcher's Broom.* 3, 4. Rare: in an orchard hedge, Alburgh, A.

ASPARAGUS, L.

*A. officinalis, L (var. hortensis). *Asparagus.* 6—8. Established in the hedge-bank of a cultivated field on Beacon Hill, Shotford, Norfolk, in the last century the site of a hall and its gardens, D. *Cf.* Introd., p. 29, note.

CONVALLARIA, L.

*C. majalis, L. *Lily of the Valley.* 5, 6. Gawdy Hall Great Wood (Mrs. Sancroft Holmes). Scarcely spreading, F.

ALLIUM, L.

A. vineale, L. *Crow Garlic.* 6, 7. Rare: bank of footpath from Harleston to the Dove House, E, G. Field between the Mendham Long Lane and White House (Mrs. Pemberton), F. Very scarce in Norfolk.

A. ursinum, L. *Broad-leaved Garlic.* 5—7. Not infrequent: Homersfield Wood; Spring Wood, Weybread; Bath Hills. Popular name *Ramsons.*

MUSCARI, Tour.

M. racemosum, Mill. *Grape-Hyacinth.* 5. Rare: on dry banks: roadside opposite Shotford Hall, E. Harleston Common, C. A doubtful native. Root poisonous.

SCILLA, L.

S. nutans, Sm. (Hyacinthus non-scriptus, L.). *Wood Hyacinth.* 4—6. Abundant in some woods; Homersfield, Weybread, Needham, etc. Occasionally with white flowers. Root poisonous. Popular name *Bluebell.*

ORNITHOGALUM, L.

*O. umbellatum, L. *Common Star of Bethlehem.* 5, 6. Not uncommon in meadows: near Spurketts' Lane by Harleston, D. Field near Potash Driftway, Weybread, E. Meadow near Flixton Hollow, I.

O. pyrenaicum, L. *Spiked Star of Bethlehem.* 6, 7. Very rare: in the Rectory Paddock (Miss Perowne, 1883), F. A casual. *Cf.* Introd., p. 26.

LILIUM, L.

*L. Martagon, L. *Turk's Cap Lily.* 6, 7. Not common, but well established. Denton Plantations, and in hedges in the neighbourhood, F. Copse near Flixton Village, G. *Cf.* Introd., p. 26.

FRITILLARIA, Tour.

F. Meleagris, L. *Common Fritillary.* 4, 5. Not infrequent in pastures. "In a field called the Seven Acres, and the adjoining ones by the side of Mendham Long Lane, near Harleston, where it also varies with a white flower" (BG, 1805), C. Meadows at Laxfield, near the church (BG). Plentiful in a field at St. Margaret's, where it has grown from time immemorial; also at St. Cross (EAH). Meadows at Luck's Mill, Needham, D. Metfield.

TULIPA, Tour.

T. sylvestris, L. *Wild Tulip.* 4. Very rare: above Weybread Watermill, Suffolk (NBG), F. It rarely flowers in its natural locality.

COLCHICUM, L.

C. autumnale, L. *Meadow Saffron.* 8, 9. Not common: in a meadow at Fressingfield (the late J. Muskett), D. Plentiful in the parish of Pulham S. Mary Magdalene (Rev. Spencer Fellows). St. Cross (EAH, 1864). Laxfield (BG). Reported also from Starston.

PARIS, L.

P. quadrifolia, L. *Herb Paris.* 5, 6. Very rare: Bedingham Wood and Tindall Wood, Ditchingham (T).

JUNCACEÆ.

JUNCUS, L.

J. bufonius, L. *Toad Rush.* 7, 8. Common in muddy and marshy places.

J. glaucus, Ehr. *Hard Rush.* 6, 7. Common by roadsides and in poor pastures.

J. diffusus, Hop. *Diffuse Rush.* 6, 7. Rare: damp places, Flixton Woods, F. St. Margaret's (EAH). *Cf.* Introd., p. 26.

J. effusus, L. *Soft Rush.* 6—8. Common in moist places.

J. conglomeratus, L. *Common Rush.* 7, 8. Common in marshy places, especially in woods.

J. obtusiflorus, Ehr. *Blunt-flowered Rush.* 7, 8. Not infrequent in wet meadows: Needham, Shimpling, etc.

J. lamprocarpus, Ehr. *Shiny-fruited Rush.* 7, 8. Abundant in the marshy meadows of the Waveney Valley. Gawdy Hall Wood; Spring Wood, Weybread.

J. acutiflorus, Ehr. *Sharp-flowered Rush.* 6—8. Not uncommon: wet ground, Rushall Wood and Spring Wood, Weybread, F. Marsh near Earsham Hall, F. Brockdish and St. Margaret's (EAH).

LUZULA, DC.

L. pilosa, Willd. *Hairy Wood Rush.* 3—5. Abundant in woods and shady places: Gawdy Hall; Shotford Hill; Homersfield, etc.

L. campestris, DC. *Field Wood Rush.* 4, 5. Common in pastures and on grassy banks.

L. multiflora, Lej. *Many-headed Wood Rush.* 6. Rare: Fir Cover, Brockdish, F.

TYPHACEÆ.

TYPHA, L.

T. latifolia, L. *Greater Reed-Mace.* 7, 8. Frequent in ponds and ditches: Baker's Barn Brickyard, Redenhall; near Mendham Priory; Dickleburgh; Harleston; Wortwell; Pulham; Shimpling, etc. Popular names *Cat's-tail, Bulrush.*

T. angustifolia, L. *Lesser Reed-Mace.* 6, 7. Not infrequent in ditches and pools: The Little Barn, Gawdy Hall North Lodge; riverside near Mendham Targets; Brockdish; Shimpling; Pulham Market; Earsham.

SPARGANIUM, Tour.

S. ramosum, Curt. *Branched Bur-reed.* 6, 7. Abundant in the marsh ditches and streams.

S. neglectum, Beeby. *Sharp-fruited Branched Bur-reed.* 7. Banks of the Waveney at Needham, and in the Weybread Beck, K. This is a new plant, and has only recently been named. *Cf.* Introd., p. 26.

S. simplex, Huds. *Unbranched Bur-reed.* 7. Frequent in ditches and streams: Redenhall Gate House; Wortwell; Dickleburgh; Brockdish; Scole; Pulham Mary, etc.

AROIDEÆ.

ARUM, L.

A. maculatum, L. *Common Cuckoo-pint.* 4, 5. Common in hedge-banks and shady places. Occasionally with a yellow spadix. Poisonous. Popular name *Lords and Ladies.*

ACORUS, L.

A. Calamus, L. *Sweet Flag.* 6. Now rare: waterside, Hoxne (JC). St. Margaret's Moat (introduced, EAH). The Waveney, near Bungay (WA). It still grows at Bungay Staithe (Mr. F. Spalding). Till very lately the floor of Norwich Cathedral was strewn with this sweet-smelling rush on certain festivals according to the general custom of the Middle Ages.

LEMNACEÆ.

LEMNA, L.

L. trisulca, L. *Ivy-leaved Duck-weed.* 6, 7. Frequent in ponds and slow ditches: The Wilderness, Harleston; Shotford Dykes; Gawdy Hall Wood, etc.

L. minor, L. *Lesser Duck-weed.* 6, 7. Common in pools and ditches.

L. gibba, L. *Gibbous Duck-weed.* 6—8. Not uncommon: in the beck at Redenhall, F. Ditch near Wortwell Low Street, CD.

L. polyrrhiza, L. *Greater Duck-weed.* Flowerless in England. Frequent: pond near Harleston Common; ditch near Alburgh Wood; Shotford Bridge Dykes; Dickleburgh.

ALISMACEÆ.

ALISMA, L.

A. Plantago, L. *Greater Water Plantain.* 7—9. Common by the side of streams and in ditches and ponds.

A. ranunculoides, L. *Lesser Water Plantain.* 6—8. Not common: marshy ground near Wingfield Castle, C. Needham Alder Carr Meadow, F. Scole and Hoxne (JC).

SAGITTARIA, L.

S. sagittifolia, L. *Common Arrow-head.* 7—9. Abundant in the Waveney and its ditches.

BUTOMUS, Tour.

B. umbellatus, L. *Flowering Rush.* 6, 7. Not infrequent: Shotford Dykes; Mendham Mill; Flixton Marshes; Stow Fen, Earsham; Dickleburgh; Scole; Shimpling; Hoxne.

NAIADACEÆ.

TRIGLOCHIN, L.

T. palustre, L. *Marsh Arrow-grass.* 7, 8. Frequent: marshes of the Waveney at Hoxne, Needham and Weybread. Boggy ground, St. Margaret's (EAH).

POTAMOGETON, L.

P. natans, L. *Floating Pondweed.* 5—7. Common in ponds, ditches, and streams.

P. rufescens, Sch. *Ruddy Pondweed.* 6, 7. Not common: Dickleburgh Moor; pond near Rushall Wood; Shimpling, F. Billingford, Pulham Mary (T). River near the Bridge at Scole (FB).

P. heterophyllus, Sch. *Different-leaved Pondweed.* 6, 7. Rare: slow water, Dickleburgh (TS).

P. lucens, L. *Shining Pondweed.* 6, 7. Common in the Waveney and its ditches.

P. prælongus, Wulf. *Long-stalked Pondweed.* 6, 7. Not common: in the Waveney at Homersfield (EAH) and above Shotford Bridge, F.

P. perfoliatus, L. *Perfoliate Pondweed.* 7. Abundant in the Waveney at Flixton and Earsham, otherwise rare, F.

P. crispus, L. *Curled Pondweed.* 6, 7. Common in the marsh dykes and ponds: var. **serratus**, Wortwell Marshes, D.

P. densus, L. *Opposite-leaved Pondweed.* 6, 7. Frequent in the Waveney and its ditches: Mendham; Wortwell; Shotford, etc.

P. pusillus, L. *Small Pondweed.* 6, 7. Frequent in ditches of the Waveney: Homersfield; Mendham, etc.

P. trichoides, Cham. *Hair-like Pondweed.* 7—10. Rare: Brockdish Dykes (EAH). Slow water ditch near Tivetshall Station; Pulham Mary; Alburgh (T). Discovered in Norfolk and added to the British Flora by the Rev. Kirby Trimmer.

P. pectinatus, L. *Fan-like Pondweed.* 6, 7. Rare: in the Waveney at Earsham, F. Ditchingham (TS).

ZANNICHELLIA, L.

Z. palustris, L. *Common Horned Pondweed.* 5—8. Frequent in the marsh ditches: Mendham Old Priory, etc.; Wortwell Broadwash.

Z. pedunculata, Reich. *Long-horned Pondweed.* 5—8. Rare: slow ditches, Gissing (T).

CYPERACEÆ.

ELEOCHARIS, R.Br.

†E. acicularis, Sm. (Scirpus acicularis, L.). *Slender Spikerush.* 7, 8. Very rare: on Stow Fen, at Earsham (T. and EAH).

E. palustris, R.Br. (Scirpus palustris, L.). *Creeping Spikerush.* 6, 7. Abundant in the marshes (Mendham, Needham, etc). Pond, Harleston Common.

E. multicaulis, Sm. (Scirpus multicaulis, Sm.). *Many-stemmed Spike-rush.* 6, 7. Rare: Needham Alder Carr Meadow; Wortwell Marshes, E. Stow Fen, Earsham (EAH).

SCIRPUS, L.

S. setaceus, L. *Bristly Club-rush.* 6—8. Rare: moist places about Earsham and Bungay (EAH).

S. lacustris, L. *Common Bull-rush.* 7, 8. Common by the riverside and in ditches.

S. sylvaticus, L. *Wood Club-rush.* 7. Frequent: banks of the Weybread Beck, C; bushy places near Foxburrows Plantation, Weybread, K; near Mendham Targets, F. Above the Mill at Syleham, and at the entrance of the St. Margaret's rivulet into the Waveney (EAH); moist woods, Ditchingham (FB).

S. Caricis, Retz. (Blysmus compressus, Panz.). *Broad-leaved Blysmus.* 7, 8. Rare: damp pastures, Shimpling, K. Ditchingham Bath Hills (T).

ERIOPHORUM, L.

E. angustifolium, Roth. *Common Cotton Grass.* 5, 6. Rare: boggy meadow, St. Margaret's (EAH), and Wortwell Marshes, E.

CAREX, L.

C. disticha, Huds. (C. intermedia, Good.). *Soft Brown Sedge.* 5, 6. Not common: pondside, Harleston Common, F; bank of the stream between St. Peter's and St. Margaret's (EAH), F.

†C. teretiuscula, Good. *Lesser Panicled Sedge.* 5, 6. Rare: marshes at Earsham and Ditchingham (BG).

C. paniculata, L. *Greater Panicled Sedge.* 6. Not common: pondside, Gawdy Hall Great Wood (EAH), and Flixton Long Plantation, F.

C. vulpina, L. *Great Spiked Sedge.* 5, 6. Abundant by ponds and streams: Baker's Barn, Redenhall; Redenhall Beck; Needham; Tivetshall; Wacton, etc.

C. muricata, L. *Greater Prickly Sedge.* 5, 6. Frequent in woods, pastures, and on banks: the Wilderness copse, Harleston; Gawdy Hall; Redenhall Road; Flixton; St. Margaret's, etc.

C. divulsa, Good. *Grey Sedge.* 5, 6. Frequent in moist shady places: roadside, Baker's Barn; Gawdy Hall Wood; Harleston; Needham; St. Margaret's, etc.

C. remota, L. *Distant-spiked Sedge.* 5, 6. Abundant in damp woods and hedge-banks; Gawdy Hall Wood; Lush Bush; Tumbrill Hill, Needham; St. Margaret's; Earsham; Hardwick; Wacton; Shimpling; Dickleburgh, etc.

C. axillaris, Good. *Axillary Sedge.* 6. Frequent in wet ditch banks: Harleston Green Lane; St. Cross; St. Margaret's; Metfield; Earsham; Pulham; Tumbrill Hill, Needham; Rushall Road, etc.

C. ovalis, Good. *Oval-spiked Sedge.* 6. Not common: marshy ground, Gawdy Hall Great Wood, F.

C. stricta, Good. *Tufted Sedge.* 4—6. Rare: wet meadows, Earsham (NBG).

C. acuta, L. *Slender-spiked Sedge.* 5, 6. Not common: wet meadows, St. Margaret's (EAH). Var. **gracilescens**, north bank of the Waveney above Syleham Mill, F. *Cf.* Introd., p. 26.

C. Goodenowii, JG. (C. vulgaris, Fr.). *Common Sedge.* 5, 6. Frequent in wet pastures and woods: Weybread Meadows; Fir Cover, Brockdish; Dickleburgh; Tivetshall; Shimpling; St. Margaret's.

C. glauca, Murr. *Glaucous Heath Sedge.* 5, 6. Common in poor pastures and damp places.

C. pilulifera, L. *Round-headed Sedge.* 5, 6. Rare: wet pastures near the Bath Hills, Ditchingham (T).

C. præcox, Jacq. *Vernal Sedge.* 4, 5. Not common: dry pastures, Harleston, F. St. Margaret's (EAH).

C. pallescens, L. *Pale Sedge.* 5, 6. Rare: damp places, Gawdy Hall Great Wood, F.

C. panicea, L. *Pink-leaved Sedge.* 6. Not common: Fir Cover, Brockdish, F. St. Margaret's Marsh, (EAH), F.

C. sylvatica, Huds. *Pendulous Wood Sedge.* 5, 6. Common in shady places: Gawdy Hall Wood; Denton Plantations; Rushall Wood; Spring Wood, Weybread; Mendham Grove, etc.

C. flava, L. *Yellow Sedge.* 5, 6. Rare: marshy ground, Fir Cover, Brockdish, F. Var. **lepidocarpa.**

C. hirta, L. *Hairy Sedge.* 5, 6. Common in meadows and wet places: Ant Hill Farm, Redenhall; Spring Wood, Weybread; meadows of the Waveney Valley, etc.

C. Pseudo-cyperus, L. *Cyperus-like Sedge.* 6. Frequent: Gawdy Hall Great Wood; Wingfield Castle Moat; Rushall; pond, Spurkett's Lane, Harleston; Weybread, etc.

C. paludosa, Good. *Lesser Pond Sedge.* 5. Abundant by the riverside: Needham Alder Carr; Flixton Long Plantation, etc.

C. riparia, Curt. *Greater Pond Sedge.* 5. Common by the riverside and in ditches.

C. rostrata, St. (**C. ampullacea**, Good.). *Bottle Sedge.* 5, 6. Not common: ditches near Needham Alder Carr; Stow Fen above Earsham Mill, F. St. Margaret's (EAH).

C. vesicaria, L. *Bladder Sedge.* 5, 6. Rare: Gawdy Hall Great Wood (EAH), F. St. Margaret's Beck (EAH).

GRAMINEÆ.

PANICUM, L.

*****P. glabrum**, Gaud. (**Digitaria humifusa**, Pers.). *Glabrous Finger-grass.* 7, 8. Rare: sandy fields by Bungay (NBG). Earsham (EAH).

SETARIA, Beauv.

*****S. viridis**, Beauv. *Green Bristle Grass.* 7, 8. Rare: Homersfield Allotments (EAH), F. Very abundant at Ditchingham (T).

*****S. glauca**, Beauv. *Glaucous Bristle Grass.* 9. Extremely abundant in sandy fields at Ditchingham (T).

PHALARIS, L.

*****P. canariensis**, L. *Canary Grass.* 7, 8. Waste ground, Needham, A.

P. arundinacea, L. (**Digraphis arundinacea**, Trin.). *Reed Canary Grass.* 7, 8. Common in streams and ditches of the Waveney.

ANTHOXANTHUM, L.

A. odoratum, L. *Sweet-scented Vernal Grass.* 5—7. Common in meadows and pastures. It gives the scent to hay.

ALOPECURUS, L.

A. agrestis, L. *Slender Fox-tail Grass.* 7, 8. Abundant in fields and waste places.

A. fulvus, Sm. *Orange-spiked Fox-tail Grass.* 6—8. Rare: in ditches on Dickleburgh Moor, F.

A. geniculatus, L. *Bent Fox-tail Grass.* 6, 7. Not common: damp meadow, Jay's Green, Harleston, A. Baker's Barn Brickyard, F. Near Mendham Priory Ruin, D. Pulham (T), etc.

A. pratensis, L. *Meadow Fox-tail Grass.* 5, 6. Common in meadows and pastures.

MILIUM, L.

M. effusum, L. *Wood Millet-grass.* 6, 7. Rare: Gawdy Hall Great Wood (EAH).

PHLEUM, L.

P. pratense, L. *Timothy Grass.* 5—10. Common in meadows and by waysides. Var. **nodosum**, roadside below Homersfield Church, F.

AGROSTIS, L.

A. canina, L. *Brown Bent Grass.* 6, 7. Not common: near Weybread House, E. Potter's Pits, Weybread, F.

A. alba, L. *Marsh Bent Grass.* Common in the marsh meadows.

A. vulgaris, With. *Fine Bent Grass.* 7, 8. Very common in cultivated and waste ground.

CALAMAGROSTIS, Ad.

C. epigeios, Roth. *Wood Small Reed.* 7. Rare: Gawdy Hall Great Wood (Mr. Flint), F. Spinney in Rushall, near Brockdish Hall (EAH), F. Earsham Wood (WA).

C. lanceolata, Roth. *Purple Small Reed.* 6, 7. Rare: pondsides, Gawdy Hall Great Wood (EAH). Blake's Grove, Gawdy Hall (BG). Earsham Wood (T).

APERA, Ad.

A. interrupta, Beauv. (**Agrostis interrupta**, L.). *Short-panicled Silky Bent Grass.* 6, 7. Very rare: in cultivated ground between Homersfield and St. Cross, 1884, A.

AIRA, L.

A. caryophyllea, L. *Silvery Hair Grass.* 6, 7. Frequent in dry places: Homersfield Heath; Mendham Pit on Withersdale Road, etc.

A. præcox, L. *Early Hair Grass.* 4—6. Abundant in dry gravelly places: pasture near Homersfield Heath; Wortwell Railway Cutting, etc.

CORYNEPHORUS, Beauv.

C. canescens, Beauv. (**Aira canescens**, L.). *Grey Hair-grass.* 6, 7. Very rare: plentiful on Homersfield Heath, D. *Cf.* Introd., p. 31.

DESCHAMPSIA, Beauv.

D. cæspitosa, Beauv. (**Aira cæspitosa**, L.). *Tufted Hair-grass.* 6—8. Common in woods and moist shady groves.

HOLCUS, L.

H. mollis, L. *Creeping Soft Grass.* 6—8. Frequent: Flixton Long Plantation; Homersfield Village; Mendham Priory Plantations; Gawdy Hall Wood; St. Margaret's.

H. lanatus, L. *Meadow Soft Grass.* 6, 7. Common in fields and by roadsides.

TRISETUM, Pers.

T. flavescens, Beauv. (**Avena flavescens**, L.). *Yellow Oat Grass.* 7, 8. Common on dry banks and in pastures.

AVENA, L.

A. pubescens, Huds. *Downy Oat Grass.* 6, 7. Not common: Gawdy Hall, near Redenhall Lodge, F. Near the Beck Bridge at St. Cross (EAH), F.

A. pratensis, L. *Narrow-leaved Oat Grass.* 6, 7. Dry pastures in the neighbourhood of St. Margaret's (EAH).

*A. strigosa, Sch. *Black Oat Grass.* 6, 7. Rare: cultivated fields, Brockdish (EAH).

A. fatua, L. *Wild Oat Grass.* 6, 7. Not common: corn-fields near Harleston, A. St. Margaret's (EAH).

ARRENATHERUM, Beauv.

A. avenaceum, Beauv. (Avena elatior, L.). *False Oat Grass.* 6, 7. Common in hedge-banks and bushy places. Var. nodosum, hedge-banks near Weybread House, F.

SIEGLINGIA, Bern.

S. decumbens, Bern. (Triodia decumbens, Beauv.). *Decumbent Heath Grass.* 7. Rare: barren pastures in the neighbourhood of St. Margaret's (EAH).

PHRAGMITES, Trin.

P. communis, Trin. (Arundo Phragmites, L.). *Common Reed.* 7, 8. Abundant in the Waveney. Pond, London Road, Harleston; Denton, etc.

CYNOSURUS, L.

C. cristatus, L. *Crested Dog's-tail Grass.* 5—8. Common in meadows and pastures.

KOELERIA, Pers.

K. cristata, Pers. *Crested Hair Grass.* 6, 7. Not common: dry pastures in the neighbourhood of St. Margaret's (EAH). Plentiful on Bungay Common, F.

CATABROSA, L.

C. aquatica, Beauv. *Water Whorl Grass.* 5—7. Frequent in damp places: Shotford Marshes; Mendham Old Priory; Needham and Brockdish Road; Pulham Market; Scole.

MELICA, L.

M. uniflora, Retz. *Wood Melic Grass.* 5—7. Frequent in shady places: The Wilderness Copse, Harleston: Gawdy Hall Wood; Mendham Hill, Norfolk; Brockdish, etc.

DACTYLIS, L.

D. glomerata, L. *Rough Cock's-foot Grass.* 6—8. Common in meadows and by roadsides.

BRIZA, L.

B. media, L. *Common Quaking Grass.* 6. Plentiful in meadows: Harleston; Starston; Needham, etc.

POA, L.

P. annua, L. *Annual Meadow Grass.* 3—10. Common in fields and waste places.

P. nemoralis, L. *Wood Meadow Grass.* 6, 7. Not common: roadside between Harleston and the Dove House, A. Gawdy Hall Great Wood, and Redenhall Green Lane, F. Nichols' Hill, Mendham: Flixton, E.

P. compressa, L. *Flat-stemmed Meadow Grass.* 6—9. Frequent in dry barren ground; roadside, Redenhall; pastures, Mendham, Scole, etc.

P. pratensis, L. *Smooth Meadow Grass.* 6, 7. Common in meadows and on hedge-banks.

P. trivialis, L. *Rough Meadow Grass.* 6, 7. Frequent in meadows: Lush Bush; Weybread; Redenhall, etc.

GLYCERIA, R. Br.

G. fluitans, R. Br. *Floating Meadow Grass.* 6—8. Common in ditches and ponds.

G. plicata, Fr. *Folded-leaved Meadow Grass.* 6—8. Rare: wet places, St. Margaret's and St. Cross (EAH).

G. aquatica, Sm. *Reed Meadow Grass.* 7, 8. Frequent by the side of streams and in ditches: Wortwell; Mendham; Needham; Pulham Market; Scole, etc.

FESTUCA, L.

F. rigida, Kunth. (Sclerochloa rigida, Link.). *Hard Meadow Grass.* 6. Frequent on dry banks: St. Cross School-ground (EAH), F. Needham Alder Carr Pit; Flixton New Road, F.

F. myurus, L. (F. Pseudo-myurus, Soy.). *Mouse-tail Fescue Grass.* 6, 7. Not common: abundant in Mendham Pit on the Withersdale Road, F.

F. sciuroides, Roth. (F. bromoides, Sm.). *Barren Fescue Grass.* 6—8. Frequent in dry places: Needham Alder Carr Pit: field near Starston Bridge; gravel pit, Earsham Station, F.

F. ovina, L. *Sheep's Fescue Grass.* 6, 7. Frequent in dry places: Homersfield Heath and roadside below the Church; gravel pit, Earsham Station, F. St. Margaret's (EAH).

F. rubra, L. *Red Fescue Grass.* 6, 7. Damp shady places: Harleston Green Lane; Flixton Park, A.

F. fallax, Th. (F. duriuscula, L). *Hard Fescue Grass.* 6, 7. Frequent in dry pastures and on banks: Mendham Pit on Withersdale Road; Rushall Road; Gawdy Hall, etc.

F. elatior, L. *Tall Fescue Grass.* 6, 7. Not infrequent: damp situations, Harleston; Flixton Park, A. Between Gawdy Hall and Alburgh, F.

  Var. **loliacea**, Huds. Frequent: meadows, Lush Bush; Weybread; St. Margaret's, etc.

  Var. **pratensis**. *Meadow Fescue Grass.* Common in pastures.

BROMUS, L

B. giganteus, L. (Festuca gigantea, Vill.). *Tall Brome Grass.* 7, 8. Not uncommon: Gawdy Hall Wood; damp places near Wortwell Mill; Mendham Mill; Flixton.

B. asper, Murr. *Rough Brome Grass.* 6, 7. Common in hedge-banks and bushy places.

B. sterilis, L. *Barren Brome Grass.* 6. Common on dry banks, walls, and in fields.

B. secalinus, L. (Serrafalcus secalinus, Bab.). *Rye Brome Grass.* 6—8. Rare: cultivated fields in the Norfolk neighbourhood of Bungay (NBG). Local name *Drauk.*

B. racemosus, L. (Serrafalcus racemosus, Parl.). *Racemose Brome Grass.* 6, 7. Rare: damp pastures, St. Margaret's (EAH).

B. commutatus, Sch. (**Serrafalcus commutatus**, Bab.). *Confused Brome Grass.* 6, 7. Frequent: pastures and cultivated ground: Mendham Mill; Needham; Rushall; Gawdy Hall; Wortwell; St. Margaret's.

B. mollis, L. (**Serrafalcus mollis**, Parl.). *Soft Brome Grass.* 6. Common on banks and in pastures.

*B. arvensis, L. (**Serrafalcus arvensis**, Godr.). *Field Brome Grass.* 7, 8. Waste places and fields; Shotford Hill; Flixton, A. Formerly at Earsham (BG).

BRACHYPODIUM, Beauv.

B. sylvaticum, Sch. *Wood False Brome Grass.* 7, 8. Plentiful in woods and shady places.

B. pinnatum, Beauv. *Barren False Brome Grass.* 7. Rare: dry open fields, Earsham (T).

LOLIUM, L.

L. perenne, L. *Common Rye Grass.* 5—7. Common in meadows and waste places. Var. *italicum occasionally as an escape from cultivation. Monstrosities are frequent. Popular name *Tinker-Tailor*.

†L. temulentum, L. *Darnel.* 6—8. A very troublesome weed among wheat in Norfolk and Suffolk (WA). Cultivated fields, St. Margaret's, formerly plentiful, now extinct (1874, EAH). Seeds very poisonous. This is the *Tare* of the Bible.

AGROPYRUM, Gært.

A. repens, Beauv. (**Triticum repens**, L.). *Creeping Couch Grass.* 6—9. Frequent: Lush Bush; Spurkett's Lane, etc. Var. **barbatum**, field opposite the Anthill Farm on the way to Mendham, F. Spurkett's Lane, A. The awned plants appear to be all A. repens.

HORDEUM, L.

H. pratense, Huds. *Meadow Barley.* 6, 7. Frequent: Needham Alder Carr Meadows; pastures near Redenhall Grange; Lush Bush, etc.

H. murinum, L. *Wall Barley.* 6, 7. Common by waysides and in waste ground.

FLOWERLESS PLANTS.
ACROGENES.

FILICES.

PTERIS, L.

P. aquilina, L. *Common Brake.* 6, 7. Common: roadsides and waste places in a sandy or gravelly soil.

ASPLENIUM, L.

A. Adiantum-nigrum, L. *Black Spleenwort.* 6—9. Rare: a few plants on Dickleburgh Church wall (DC). Near Ditchingham Hall (Mr. F. Spalding).

A. viride, Huds. *Green Spleenwort.* 6—9. Very rare: discovered by Mr. T. M. Spalding, forty years ago, on an old wall between Denton and Bungay, and growing there still (Mr. F. Spalding), F. *Cf.* Introd., p. 28.

A. Trichomanes, L. *Maiden-hair Spleenwort.* 5—10. Rare: occasionally on buildings at Gawdy Hall (Mr. Flint). Bedingham Church wall (Mr. F. Spalding).

A. Ruta-muraria, L. *Wall Rue Spleenwort.* 5—9. Not infrequent: Harleston Common, B. Fressingfield Church wall. Syleham (DC). Dickleburgh (TS). Ditchingham Church wall (Mr. F. Spalding). Long Stratton Church (WA).

ATHYRIUM, Roth.

A. Filix-fœmina, Roth. *Lady Fern.* 6, 7. Rare: Flixton Long Plantation, F. Formerly in Gawdy Hall Wood (the late J. Muskett).

CETERACH, Will.

C. officinarum, Will. *Scaly Spleenwort.* 4—10. Very rare: Mendham Church wall, B. Bridge near Forncett Station (Mr. H. F. Wilson). *Cf.* Introd., p. 28.

SCOLOPENDRIUM, Sm.

S. vulgare, Sym. *Common Hart's Tongue.* 7, 8. Frequent in shady places.

CYSTOPTERIS, Bern.

C. fragilis, Bern. *Brittle Bladder Fern.* 6, 7. On an old wall at Harleston (1843, the late Mr. Muskett and Mr. T. M. Spalding); locality destroyed. Harleston Station wall (1884, B.). *Cf.* Introd., p. 28.

POLYSTICHUM, Roth.

P. lobatum, Pres. (Aspidium lobatum, Sw.). *Prickly Shield Fern.* 7, 8. Shady hedge-bank near Redenhall Gate-house, B. Gissing (T).

Var. aculeatum, Sym. Not infrequent: Redenhall, Rushall, Needham, etc.

P. angulare, Pres. (Aspidium angulare, Willd.). *Angular Shield Fern.* 7, 8. Rare: shady hedge-banks, Shelton (Mrs. Sancroft Holmes). Formerly frequent at Dickleburgh (DC).

LASTRÆA, Pres.

L. Filix-mas, Pres. (Nephrodium Filix-mas, Rich.). *Male Buckler Fern.* 6, 7. Generally distributed in woods and hedge-banks.

L. spinulosa, Pres. (Nephrodium spinulosum, Desv.). *Spinulose Buckler Fern.* 7—9. Rare: moist places in several parts of Gawdy Hall Great Wood (Mr. Flint), F.

L. dilatata, Pres. (Nephrodium dilatatum, Desv.). *Broad Buckler Fern.* 7—9. Very rare: sparingly in a shady lane at Shelton (Mrs. Sancroft Holmes).

POLYPODIUM, L.

P. vulgare, L. *Common Polypody.* 7—11. Common in banks and on trees.

OPHIOGLOSSUM, L.

O. vulgatum, L. *Adder's Tongue.* 5, 6. Not infrequent: pasture near Jay's Green, Harleston, B. Gawdy Hall Great Wood, F. Meadow, St. Cross, G. Gawdy Hall Lawns (Mrs. Pemberton), F. Fressingfield, D.

EQUISETACEÆ.

EQUISETUM, L.

E. **arvense,** L. *Common Horse-tail.* 4. Common in cornfields and gravel-pits.

E. **palustre,** L. *Marsh Horse-tail.* 6, 7. Frequent in ditches: Needham Alder Carr Meadows; Weybread, etc.

E. **limosum,** Sm. *Smooth Horse-tail.* 6, 7. Common in ditches and streams.

E. **hyemale,** L. *Rough Horse-tail.* 7, 8. Very rare: "I believe I have seen it growing at the Earsham end of the Bath Hills" (Mr. F. Spalding). Hedenham (WA).

CHARACEÆ.

CHARA, L.

C. **fragilis,** Desv. *Slender Chara.* 6—8. Pond, Harleston Common, F.

C. **aspera,** Willd. *Rough Chara.* 8. Ditches near Weybread Mill, Suffolk, F.

C. **vulgaris,** L. (Chara fœtida, A. Br.). *Common Chara.* 6—8. Abundant in the marsh dykes: Shotford; Mendham Priory Meadows, etc., F.

APPENDIX.

ADDITIONAL SPECIES RECORDED WITHIN EIGHT MILES.

Viola canina, L. (Var. flavicornis.) *Dillenius' Dog Violet.* 4, 5. Sandy places, Broome Heath (T).

Sagina nodosa, Mey. *Knotted Spurrey.* 7, 8. Broome Fen (T).

Ulex nanus, Forst. *Dwarf Furze.* 8—11. Stuston ; Bungay.

Medicago maculata, Sib. *Spotted Medick.* 5—8. Bungay (NBG).

Trifolium suffocatum, L. *Suffocated Clover.* 6, 7. Bungay Common (EAH) ; Broome (T). *Cf.* Introd., p. 31.

*Sedum reflexum, L. *Crooked Stone-crop.* 6, 7. Bungay, E.

Epilobium angustifolium, L. *Rose-bay Willow-herb.* 6—8. Hedenham (T).

Epilobium obscurum, Sch. *Short-podded Willow-herb.* 7, 8. Bungay (Stock).

Apium graveolens, L. *Celery.* 6—8. Bungay (NBG). *Cf.* Introd., p. 30.

*Inula Helenium, L. *Elecampane.* 7, 8. Mettingham (WA).

Taraxacum palustre (DC). *Dandelion.* 5, 6. Damp meadows, Broome (WA).

*Mimulus luteus, L. *Monkey-flower.* 7—9. Bungay (HS).

Veronica triphyllos, L. *Fingered Speedwell.* 4. Bungay (Suckling).

Pedicularis palustris, L. *Marsh Lousewort.* 5—9. Bungay (Rev. W. M. Hind).

Calamintha arvensis, Lam. (C. Acinos, Clair.). *Basil Thyme.* 6—9. Gravelly places, Bungay (Stock). Brooke (TS).

Chenopodium urbicum, L. *Upright Goosefoot.* 7, 8. Bungay (HS).

Chenopodium rubrum, L. *Red Goosefoot.* 7, 8. Bungay (HS).

Rumex maritimus, L. *Golden Dock.* 7, 8. Bungay (NBG).

Rumex palustris, L. *Marsh Dock.* 8—10. Bungay (NBG).

***Aristolochia Clematitis,** L. *Birthwort.* 6—9. Stuston (Dr. Amyot).

***Urtica pilulifera,** L. *Roman Nettle.* 6—8. Bungay (FB).

Salix pentandra, L. *Bay-leaved Willow.* 4, 5. Bungay (WA).

Salix aurita, L. *Round-eared Willow.* 5. Bungay (WA).

Epipactis purpurata, Sm. *Purple Helleborine.* 7, 8. Hedenham (WA).

Epipactis palustris, Cr. *Marsh Helleborine.* 7, 8. Broome Fen (T).

Gagea fascicularis, Sal. *Yellow Star of Bethlehem.* 4. Pastures, Shipmeadow (WA). Confirmed by the Rev. E. A. Holmes in 1884.

Juncus squarrosus, L. *Heath Rush.* 6, 7. Bungay (HS).

Potamogeton plantagineus, Ducr. *Plantain-leaved Pondweed.* 6, 7. In ditches by Broome (T).

Scirpus pauciflorus, Light. *Few-flowered Rush.* 7, 8. Broome Fen (T).

Schœnus nigricans, L. *Black Bog Rush.* 6, 7. Broome Fen (T).

Cladium germanicum, Sch. *Prickly Twig Rush.* 7, 8. Bungay (WA).

Carex dioica, L. *Creeping Diœcious Sedge.* 5, 6. Broome (BG).

Carex pulicaris, L. *Flea Sedge.* 5, 6. Bungay (WA).

Carex echinata, Murr. (C. stellulata, Good.). *Little Prickly Sedge.* 5, 6. Bungay (Stock).

Carex strigosa, Huds. *Loose Pendulous Sedge.* 5, 6. Hedenham (BG).

Carex distans, L. *Loose Sedge.* 6. Bungay (HS).

Carex Œderi, Ehr. *Œder's Sedge.* 6, 7. Bungay (Stock).

*****Panicum sanguinale,** Scop. *Hairy Finger Grass.* 8. Broome (Stock).

Bromus erectus, Huds. *Upright Brome Grass.* 6, 7. Bungay (Woodward).

Agropyron caninum, Beauv. *Fibrous Couch Grass.* 7. Bungay (HS).

Lastrea Thelypteris, Pres. *Marsh Buckler Fern.* 7, 8. Bungay (WA).

Equisetum maximum, Lam. *Great Horsetail.* 4. Bungay (HS).

Lycopodium inundatum, L. *Common Club Moss.* 8, 9. Bungay (HS).

V.
OBSERVATIONS ON THE BIRDS.

OBSERVATIONS ON

THE BIRDS OF THE HARLESTON DISTRICT.

BY

CHARLES CANDLER.

THE following list, compiled for the most part from casual notes entered in a diary from time to time during the last seven years, without any thought of publication, must not be taken as a complete catalogue of the birds of the district. Many species which I have strong reasons to believe are regular or frequent visitors to our neighbourhood are here omitted, as for want of a larger number of observers (or perhaps I should say of more careful attention on my own part) they have hitherto escaped detection. So far, however, as the observations extend, I have taken care to insure their accuracy, and I think that in this respect they may be considered reliable. It will be found that the notes relate to a much more restricted area than that covered by the observations of the Harleston Botanical Club. Indeed, of the 126 species mentioned below, upwards of 100 have been met with in the two parishes of Redenhall and Mendham alone; and, with a very few exceptions, all have occurred within three miles of Harleston Railway Station. In conformity with Messrs. Gurney and Southwell's authoritative *List of the Birds of Norfolk*, recently published,* the arrangement and nomenclature adopted by the editors of the fourth edition of Yarrell's *British Birds* have been followed; and I am personally indebted to Mr. Thomas Southwell for kindly looking through my notes and otherwise assisting me.

The species marked with an asterisk are known to breed in the district.

1. WHITE-TAILED EAGLE (*Haliæetus albicilla*).

A bird in immature plumage is still preserved at Gawdy

* Transactions of the Norfolk and Norwich Naturalists' Society, vol. iv., pp. 259—286 and 397—432.

Hall, which, according to the Journal of the Rev. Wm Whitear,† was shot in the wood on January 29th, 1823.

2. PEREGRINE FALCON (*Falco peregrinus*).

A female in second year's plumage was shot near Gawdy Hall on the 12th January, 1884.

3. *HOBBY (*Falco subbuteo*).

Rare. Mr. Stevenson records the nesting of this hawk at Thorpe Abbots. About fifteen years ago a pair of Hobbies built a nest in an oak-tree in Gawdy Hall Park. The birds were shot, and, with their nest and eggs, are preserved at the Hall. The Hobby has also been killed at Flixton (Suckling's *Suffolk*).

4. MERLIN (*Falco œsalon*).

Sir E. Kay's keeper has killed three or four Merlins at Thorpe Abbots.

5. *KESTREL (*Falco tinnunculus*).

Though a constant war is waged against this bird, it is still by no means uncommon. For several years a pair nested in the tower of Redenhall Church, and I have known the bird to breed in the fragment of old wall which is all now remaining of the ruins of Mendham Priory. (In the summer of 1886 I found a Kestrel's nest, containing six eggs hard sat upon, in the ruinous tower of Linstead Magna Church, within a few feet of the bell chimed every Sunday for service.)

6. *SPARROW-HAWK (*Accipiter nisus*).

Less common than the Kestrel, but not yet extinct as a resident with us.

7. [BUZZARD (*Buteo — ?*).

Frequent reports of large raptorial birds seen in the neighbourhood are brought to me in autumn; most probably Common or Rough-legged Buzzards on passage.]

8. *TAWNY OWL (*Strix aluco*).

A few Tawny Owls still exist in the woods around Flixton Hall, but elsewhere in the district the bird is either extinct or very rare. Two or three years ago Sir E. Kay's keeper killed one of these birds at Thorpe Abbots.

† The Rev. Wm. Whitear, M.A., F.L.S., was Rector of Starston from 1803 till his death in 1826. His Journal, covering the years 1809—1826, was published by the Norfolk and Norwich Naturalists' Society, in their "Transactions" for 1880—81, vol. iii., pp. 231—262.

9. *Long-eared Owl (*Asio otus*).

Occasionally found in the neighbourhood of the town at Shotford Hill, and in other similar localities. In the fir woods at Flixton the bird is common.

10. Short-eared Owl (*Asio accipitrinus*).

An autumn visitor, occasionally shot by our sportsmen in the stubbles and turnip-fields.

11. *Barn Owl (*Aluco flammeus*).

Too many of these most useful and interesting birds pass every year through the hands of our local bird-stuffers. They cling tenaciously to their old nesting-places. A pair have for many years haunted a hollow elm near the town, though frequently robbed of their eggs, stoned, and shot at. If this bird received the protection it deserved, there would be scarcely an old homestead in the district without its pair of Barn Owls.

12. *Red-backed Shrike (*Lanius collurio*). "Butcher Bird."

Now sparsely distributed in summer, the trim fences of the modern farm affording the bird no nesting cover.

13. *Spotted Flycatcher (*Musicapa grisola*).

A common summer visitant.

14. *Mistletoe Thrush (*Turdus viscivorus*). "Fulfer" and "Dow-fulfer."†

Common. A noisy and conspicuous bird in autumn.

15. *Song-Thrush (*Turdus musicus*). "Mavis."

Common; disappearing in seasons of severe and prolonged frost, as the winter of 1880—81.

16. Redwing (*Turdus iliacus*), and

17. Fieldfare (*Turdus pilaris*). "Fulfer."

Regular winter immigrants, the latter being the more conspicuous and better known.

18. *Blackbird (*Turdus merula*).

Common. A hardier bird than its congener the Thrush.

† In the hope of interesting my boy friends in the subject, I have given the local names of a few of those birds with which they are most familiar. The word "*fulfer*" is here spelt as it is locally pronounced. The name is also applied, and more properly belongs, to the Fieldfare.

19. Ring Ouzel (*Turdus torquatus*).

"1818.—April 25th.—A Ring Ouzel was shot this day upon Spurling's farm in this parish [Starston]."—Mr. Whitear's Journal (Trans. Norfolk and Norwich Naturalists' Society, vol. iii., p. 247).

20. *Hedge Sparrow (*Accentor modularis*).

A common and, I believe, a constant resident.

21. *Redbreast (*Erithacus rubecula*).

Common. Breeding with us from year to year in apparently unvarying numbers.

22. *Nightingale (*Daulius luscinia*).

A summer visitor, sparingly distributed.

23. *Redstart (*Ruticilla phœnicurus*).

A regular summer visitor, frequenting our gardens, though in small numbers.

24. Stonechat (*Saxicola rubicola*).

Not at all common.

25. *Whinchat (*Saxicola rubetra*).

Scarce, especially on the Norfolk side of the river.

26. Wheatear (*Saxicola œnanthe*).

Met with on migration.

27. *Reed Warbler (*Acrocephalus streperus*). "Reedbird."

A summer visitor, nesting regularly in the beds of *Arundo Phragmites* which fringe the Waveney in many places between Weybread and Needham Mills.

28. *Sedge Warbler (*Acrocephalus schœnobænus*).

A common summer visitor, haunting rank growths of vegetation near streams, ponds, and ditches all over the district.

29. Grasshopper Warbler (*Acrocephalus nævius*).

Mr. Whitear writes in his Journal, under date April 27th, 1821: "Saw a Grasshopper Warbler in the hedge of the Beck meadow." He also notes the arrival of the bird at Starston on the 23rd April, 1822.

30. *WHITETHROAT (*Sylvia rufa*). "Hayjack."

One of the most abundant of our summer migrants.

31. *LESSER WHITETHROAT (*Sylvia curruca*).

"We have noticed the Lesser Whitethroat more than once at Starston, and have also procured its eggs at the same place" (Sheppard and Whitear's *Catalogue of the Norfolk and Suffolk Birds* 1825, p. 19). Mr. F. Boyce tells me that he has often met with this bird at Redenhall, and it appears to be not rare in the neighbourhood.

32. *GARDEN WARBLER (*Sylvia hortensis*), and

33. *BLACKCAP (*Sylvia atricapilla*).

Summer migrants, generally distributed in groves and gardens.

34. WOOD WREN (*Phylloscopus sibilatrix*).

"This species is scarce. A specimen was killed at Starston." —(List of the birds of the county in Stacy's *History of Norfolk*, 1829.)†

35. *WILLOW WREN (*Phylloscopus trochilus*). "Ground-oven."

Common in orchards and plantations through the summer.

36. *CHIFFCHAFF (*Phylloscopus collybita*).

A summer visitant, and nearly as plentiful in our district as the Willow Wren.

37. *GOLDEN-CRESTED WREN (*Regulus cristatus*).

Not very common in summer; more frequently met with in autumn and winter.

38. *WREN (*Troglodytes parvulus*).

Very common.‡

---

† This list was contributed by John Hunt, of Norwich, author of an illustrated work on British Birds, and a friend and correspondent of the Rev. Wm. Whitear, from whom, no doubt, he received the above information.

‡ Some years ago, before the days of compulsory education, Wrens were familiarly known around Harleston as "stags;" and "stag-hunting"—that is, stoning a Wren up and down a hedge till a shot from one side or the other killed the bird—was a favourite sport with our boys, and even young men. The school-attendance inspector is in many ways a good friend to our wild birds!

39. *TREE CREEPER (*Certhia familiaris*).

A common resident.

40. *NUTHATCH (*Sitta cæsia*). "Nutcracker."

Fairly common. An examination of the old oaks by the roadside in Gawdy Hall Wood will show traces of the work of this bird in the form of broken nutshells firmly wedged in the crevices of the bark.

41. *GREAT TITMOUSE (*Parus major*), "Pickcheese," "Blackcap," and

42. *BLUE TITMOUSE (*Parus cœruleus*).

Two very common birds.

43. *COAL TITMOUSE (*Parus britannicus*), and

44. *MARSH TITMOUSE (*Parus palustris*).

These two species appear to be equally plentiful with us. The Marsh Tit is by no means confined to the river valley, but has been frequently observed by Mr. F. Boyce at Redenhall, and is, I believe, generally distributed in the upland districts. It has been shot in my father's garden in the town.

45. *LONG-TAILED TITMOUSE (*Acredula caudata.*) "Puddingpoke."

Tolerably common. Less frequently noticed in summer than in winter, when small parties are constantly met with actively moving about in search of food.

46. *PIED WAGTAIL (*Motacilla lugubris*). "Penny Wagtail."

Not common, only a few pairs nesting in the neighbourhood.

47. *YELLOW WAGTAIL (*Motacilla raii*). "Capering Longtail."

A summer visitant. Several pairs nest regularly along the margins of the dykes on the Mendham Marshes.

48. *TREE PIPIT (*Anthus trivialis*).

A common summer migrant. The slopes of the railway cuttings and embankments are favourite nesting haunts of this bird.

49. *MEADOW PIPIT (*Anthus pratensis*). "Titlark."

Not very common; found frequenting the rough water-meadows by the Waveney.

50. *SKYLARK (*Alauda arvensis*).

Very common.

51. *REED BUNTING (*Eberiza schœniclus*). "Blackcap."

Common in the river valley at all times of the year.

52. *BUNTING (*Emberiza miliaria*).

A common and conspicuous bird, particularly in the meadows near the town. Its nest is very rarely found.

53. *YELLOW HAMMER (*Emberiza citrinella*).

An abundant and, I think, increasing species. Nesting in banks and ditches, and feeding on grain, it has been little affected by the destruction of the old hedgerows.

54. *CHAFFINCH (*Fringilla cœlebs*). "Spink."

Abundant.

55. BRAMBLING (*Fringilla montifringilla*).

An occasional winter visitor.

56. *TREE SPARROW (*Passer montanus*).

Mr. F. Boyce has identified the Tree Sparrow at Redenhall, and has taken its eggs from a hole in a tree by the beck.

57. *HOUSE SPARROW (*Passer domesticus*).

Abundant everywhere.

58. (* ?) HAWFINCH (*Coccothraustes vulgaris*).

Not at all rare. I have a bird shot in a garden in the town in November, 1880, and some dozen others have been killed in the immediate neighbourhood during the last few years.†

59. *GREENFINCH (*Coccothraustes chloris*). "Green Linnet."

Only less abundant than the House Sparrow.

---

† As to the occasional abundance of this bird at Diss, see Mr. Southwell's note in his edition of Lubbock's *Fauna of Norfolk*, p. 63.

60. *GOLDFINCH (*Carduelis elegans*). "King Harry."

A few pairs nest in our orchards every summer, but unfortunately their broods are generally secured by some bird-fancier. The bird will soon be rare.

61. SISKIN (*Carduelis spinus*).

A winter migrant, uncertain in numbers, frequenting in severe weather the alders by the river.

62. *LESSER REDPOLL (*Linota rufescens*).

Resident, I think, in small numbers, but better known as a winter visitor. Some years ago a very large flock frequented, at this season, a wood near Mendham Mansion.

63. *LINNET (*Linota linaria*). "Brown," "grey," or "red Linnet."

Not very plentiful. Nests frequently in the furze bushes on the now enclosed and cultivated tract of land still known as "Shotford Heath."

64. *BULLFINCH (*Pyrrhula europæa*). "Blood Olph."

A resident, sparingly distributed.

65. *STARLING (*Sturnus vulgaris*).

Very common.

66. *CARRION CROW (*Corvus corone*).

A much-persecuted and now scarce bird. For years a pair has attempted to breed in a plantation near Weybread Hall. One or two nests are found every summer in the Gawdy Hall Woods. The bird is frequently seen at Thorpe Abbots, and nests in some high trees near the river.

67. HOODED CROW (*Corvus cornix*). "Kentish Crow."

A regular and common winter migrant.

68. *ROOK (*Corvus frugilegus*).

Abundant. The largest rookeries in the neighbourhood are at Flixton and Gawdy Hall. In 1881 the rooks returned again to their old quarters in the lofty elms of the "White House" garden, which had been deserted by the birds for several years. Three nests were then built, and the number has increased every year since.

69. *Jackdaw (*Corvus monedula*). "Cadder."

Common enough in winter, mingling with the rooks, or flying separately in flocks often of considerable size. A very small number of jackdaws breed in the neighbourhood. A few pairs nest in the hollow trees at Flixton Park.

70. Magpie (*Pica rustica*).

As a resident, the Magpie is quite extinct in the district under observation, and can only be included in this list as a very rare visitor. Mr. J. A. Holmes has observed the bird at Flixton within the last fifteen or twenty years, and I learn from two informants that it has been seen much more recently at Thorpe Abbots. The Rev. H. T. Frere tells me that a Magpie was seen for some weeks last year at Gissing, which, however, he thinks may have been an escaped bird, as it was very tame. The last *pair* of Magpies which Mr. Frere recollects in this part of Norfolk frequented the neighbourhood of Frenze fifty years ago.

71. *Jay (*Garrulus glandarius*).

This bird still holds its own against the keepers, thanks to its silent and wary habits during the breeding season.

72. *Swallow (*Hirundo rustica*).

Common through the summer.

73. *House Martin (*Chelidon urbica*).

Not very plentiful. Some years ago my brother counted forty-five nests under the broad eaves of a thatched house at Alburgh. This is the largest colony I have met with. The front of the Swan Hotel in this town is a well-known rendezvous of the Martins before their autumn emigration. Numbers of them may be seen here in the early morning sitting in rows along the narrow ledges of the brickwork, and clustering upon the ornamental ironwork supporting the old sign.

74. *Sand Martin (*Cotile riparia*).

Common. A considerable number nest in the large sand-pit at the foot of Needham Hill.

75. *Swift (*Cypselus apus*). "Devil" and "Deviling."

Not plentiful. A few pairs nest in the tower of Redenhall Church every year.

76. *NIGHTJAR (*Caprimulgus europæus*).

A rare bird in the vicinity of Harleston, but more frequently met with in Wortwell, Mendham, and Homersfield, on lighter and warmer soils. A young bird was shot at Brockdish last August, and another a few weeks later at Weybread. At Thorpe Abbots, in the extreme west of the district, the bird is scarce, though a wooded country and gravelly soil would seem to be favourable conditions.

77. *CUCKOO (*Cuculus canorus*).

A common summer visitor. In July, 1881, a Cuckoo laid her egg in a Spotted Flycatcher's nest, built in the cleft of a pear-tree, in an orchard near the town. The young Cuckoo was hatched, and partly reared by the Flycatchers, but unfortunately was taken from the nest before fully fledged. (A similar instance of a Cuckoo's egg having been deposited in a Flycatcher's nest, came under my notice at Fundenhall in the same year.)

78. HOOPOE (*Upupa epops*).

One killed at Harleston in April, 1859 (Stevenson's *Birds of Norfolk*).

79. ROLLER (*Coracias garrulus*).

Mr. Stevenson notes the occurrence of a Roller at Earsham.

80. *KINGFISHER (*Alcedo ispida*).

These birds suffered much from the high floods and severe frosts of the years 1878—1881, during which period several were found dead. In 1878 we found a Kingfisher's nest, containing seven eggs, in a Sand Martin's burrow in a gravel pit close to the town, and at some distance from the river. Probably not more than three or four pairs now breed in the vicinity, but in autumn a considerable immigration takes place, and the bird is then frequently seen, and much too frequently shot.

81. *GREEN WOODPECKER (*Gecinus viridis*).

The most common of our three Woodpeckers

82. *GREAT SPOTTED WOODPECKER (*Dendrocopus major*).

Less common than the last-mentioned species, but not rare in the Gawdy Hall Woods, the Starston Plantations, and other suitable localities.

83. *LESSER SPOTTED WOODPECKER (*Dendrocopus minor*).

Rare or perhaps seldom observed. A bird was shot at Mendham in February, 1881, and one at Pulham Market last October.

84. *WRYNECK (*Iynx torquilla*). "Cuckoo's mate."

A common summer visitor, and one of the best known heralds of the spring.

85. *RING DOVE (*Columba palumbus*).

Very common, nesting in nearly all our woods and plantations. In autumn great numbers resort to the oak groves to feed on the acorns, and are shot from huts made of furze and hurdles. I have known fifty-four birds, in a small plantation at Starston, fall to a single gun in one day. In the crops of these birds I have found, on dissection, an almost incredible number of entire acorns.

86. STOCK DOVE (*Columba œnas*).

Nests in the hollow trees at Gawdy Hall and Flixton (where I found, in 1880, a nest containing *three* eggs). I have also found its nest in a hole in the masonry of the ruined wall of Mendham Priory. Flocks of Stock Doves are occasionally seen here in winter.

87. *TURTLE DOVE (*Turtur communis*).

A summer visitant, nesting in the plantations at Gawdy Hall, Starston and Shotford Hill, and, indeed, wherever it can find a sufficiently dense cover.

88. *PHEASANT (*Phasianus colchicus*).

89. *PARTRIDGE (*Perdix cinerea*).

90. *RED-LEGGED PARTRIDGE (*Caccabis rufa*).

The "French" is decidedly less abundant than the "English" bird in this district.

91. QUAIL (*Coturnix communis*).

An uncertain visitor. In 1880, I believe, one or two pairs nested near the town, and their curious trisyllabic note was noticed by many persons. A bird was shot at Alburgh in the autumn of the same year. Several arrived on the Suffolk side of Scole, May, 1868.—(Babington's *Birds of Suffolk*.)

92. *LAND-RAIL (*Crex pratensis*).

Not a common bird with us, though hardly a summer passes without one or two pairs breeding in the neighbourhood.

93. SPOTTED CRAKE (*Porzana parva*).

Mr. John A. Holmes informs me that he killed a bird of this species some years ago in Gawdy Hall Wood, when pheasant-shooting.

94. *WATER-RAIL (*Rallus aquaticus*).

Rarely seen through the summer, though not uncommon. In winter it is often shot by our sportsmen along the river-side.

95. *MOORHEN (*Gallinula chloropus*).

Common. Between Shotford Bridge and Mendham Priory the Waveney, during the summer, is in many places silted up and choked from bank to bank by a rank growth of Œnanthe, Sium, and other weeds, which afford protection to numbers of waterhens.

96. COOT (*Fulica atra*).

Occasionally shot upon the river in winter.

97. RINGED PLOVER (*Ægialetis hiaticula*).

Mr. James Elsey, of Mendham, tells me that some years ago he stuffed a Ringed Plover, which was shot in a ploughed field near his house.

98. GOLDEN PLOVER (*Charadrius pluvialis*).

Flocks appear in our fields in autumn, and again in spring, when they have been shot with the black breast of their breeding plumage.

99. *LAPWING (*Vanellus vulgaris*). " Peewit."

The Lapwing frequents our fields and marshes in varying numbers throughout the winter. A few pairs breed here and there in the valley of the Waveney, and occasionally upon the larger upland fields.

100. WOODCOCK (*Scolopax rusticola*).

A regular autumn visitor, though only met with in very small numbers. At Flixton, this autumn, twelve birds have been shot in a week. In the woods at Thorpe Abbots and Brockdish more than three birds are rarely killed in one day.

101. *COMMON SNIPE (*Gallinago cœlestis*).

The Snipe appears regularly upon our marshes in autumn, its numbers, however, varying greatly with the condition of the weather. It breeds with us in, I hope, increasing numbers. Last summer I heard of four nests within a mile of Shotford Bridge.

102. JACK SNIPE (*Gallinago gallinula*). "Half Snipe."

A winter visitor, generally in very small numbers.

103. DUNLIN (*Tringa alpina*).

I have noted but one occurrence of the Dunlin near Harleston—a bird shot a few years ago by an upland pond-side in Mendham.

104. COMMON SANDPIPER (*Totanus hypoleucus*).

Often seen by the river-side in winter.

105. GREEN SANDPIPER (*Totanus ochropus*).

Single birds of this species have been seen in this neighbourhood at almost every season of the year. It is not infrequently shot by the side of marsh dykes in autumn and winter.

106. CURLEW (*Numenius arquata*).

Frequently seen or heard passing overhead.

107. LESSER TERN (*Sterna minuta*).

An occasional visitor. Two or three have been shot in Mendham.

108. BLACK TERN (*Hydrochelidon nigra*).

One shot by the river at Mendham, May, 1883.—(Babington's *Birds of Suffolk*.).

109. BLACK-HEADED GULL (*Larus ridibundus*).

A common visitor in autumn and winter.

110. COMMON GULL (*Larus canus*).

Not infrequently seen in winter.

J

111. HERRING GULL (*Larus argentatus*), and

112. LESSER BLACK-BACKED GULL (*Larus fuscus*).

Birds of these species frequently wander up the valley in winter, those shot being generally in immature plumage. A beautiful adult Herring Gull has been recently killed at Mendham.

113. KITTIWAKE GULL (*Rissa tridactyla*).

Numbers of Kittiwakes may sometimes be seen in winter at Wortwell and Homersfield when the marshes are flooded.

114. DUSKY PETREL (*Puffinus obscurus*).

Mr. Wm. Hartcup, of Bungay, has a male bird of this species, which was found dead at Earsham in the spring of 1858. This is the only example of the Dusky Petrel known to have occurred in Great Britain.†

115. STORM PETREL (*Procellaria pelagica*).

Some years ago a Storm Petrel was brought to the late Mr. James Muskett, which had fallen down the chimney of a cottage, at Clintergate, Redenhall, during a gale of wind.

116. LITTLE AUK (*Mergallus alle*).

More than sixty years ago a storm-driven wanderer of this species was caught alive in a stackyard at Pulham.—(Sheppard and Whitear's *Catalogue of the Norfolk and Suffolk Birds*, p. 60.)

117. *LITTLE GREBE (*Podicipes fluviatilis*). "Dobchick."

The Dabchick is frequently shot on the river in winter, and I think nests with us, for birds have been noticed all through the summer.

---

† Mr. H. Stevenson has given an interesting account of the history of this specimen, and its discovery by him after it had been lost sight of for 24 years, in the Transactions of the Norfolk and Norwich Naturalists' Society, vol. iii., pp. 467—473. To Mr. Stevenson we are primarily indebted for our illustration of this bird, which is reproduced with his approval from a photograph taken by Messrs. Sawyer and Bird under his direction. The Dusky Petrel has, with its Australian representative *Puffinus assimilis*, an extensive ocean range in both northern and southern hemispheres, but it has very rarely been met with north of the Mediterranean. The only other example, which has been noted near the shores of Britain, was caught alive on a vessel off the southern coast of Ireland in 1853.

118. *HERON (*Ardea cinerea*). "Harnser.'

In the spring of 1884, a party of Herons, which had haunted the vicinity of Flixton Hall during the winter, took up their quarters in a grove of lofty oaks near the Thicket Wood. Great care was taken that the birds should not be disturbed, and seven nests were built in the first year. In 1885 the number of nests fell to three, and remained the same in 1886. Last spring the little colony increased to four pairs. A single pair of Herons have, in recent years, several times nested in Gawdy Hall Wood.† A small heronry has also been established within the last seven years in the parish of Thorpe Abbots, the Herons having chosen as a breeding station a plantation on the grounds of Thorpe Hall. The number of nests has varied from year to year, but has never exceeded seven, and last year only three pairs of birds bred in the locality.

119. NIGHT HERON (*Nycticorax griseus*).

A bird of this species was shot in a fir-tree in the Vicarage Garden, at Mendham, on the 10th of May, 1879.

120. WHITE-FRONTED GOOSE (*Anser albifrons*).‡

Several of these birds were shot at Wortwell in February, 1883.

121. *WILD DUCK (*Anas boschas*).

Common. This bird breeds by the ponds in Gawdy Hall Wood, and in other suitable localities where protected.

122. SHOVELLER (*Spatula clypeata*).

I have only noted two or three of these ducks shot in winter along the Waveney.

123. TEAL (*Querquedula crecca*).

A regular winter visitant to our streams and marshes.

124. WIGEON (*Mareca penelope*). "Smee Duck."

A winter visitor.

---

† The following entry occurs in Mr. Whitear's notebook: "A pair of Herons bred three times at Gawdy Hall; the eggs were taken twice, and the young once, about the year 1808." (Trans. Norfolk and Norwich Naturalists' Society, vol. iii., p. 258.)

‡ Triangles of Geese frequently pass overhead in winter, but the birds are rarely killed, and the White-fronted Goose is the only species I have had an opportunity of examining. I might, however, with perfect safety, add the Pink-footed Goose to my list. The Rev. H. T. Frere tells me he has identified the bird at Burston.

125. POCHARD (*Fuligula ferina*).†

Occasionally shot in winter.

126. GOLDEN EYE (*Clangula glaucion*) !

Mr. J. A. Holmes tells me that some years ago he shot a Duck at Needham, which he identified with the "Morillon," of Bewick. It has been occasionally seen at Oakley, near Hoxne (Babington's *Birds of Suffolk*).

† Here, as elsewhere, the Teal, Wigeon, and Pochard, are known as "Half-Duck." My list of Wild Fowl visiting our river marshes in winter is very defective, and might be largely added to if our local gunners would pay as much attention to the birds they kill as they do to their sport.

INDEX TO THE BIRDS.

(The numbers refer to those prefixed to the names in the foregoing Observations.)

Barn Owl, 11.
Blackbird, 18.
Blackcap, 33.
Black-headed Gull, 109.
Black Tern, 108.
Blue Titmouse, 42.
Brambling, 55.
Bullfinch, 64.
Bunting, 52.
Buzzard, 7.

Carrion Crow, 66.
Chaffinch. 54.
Chiffchaff, 36.
Coal Titmouse, 43.
Coot, 96.
Cuckoo, 77.
Curlew, 106.

Dunlin, 103.
Dusky Petrel, 114.

Fieldfare, 17.

Garden Warbler, 32.
Golden-crested Wren, 37.
Golden Eye, 126.
Golden Plover, 98.
Goldfinch, 60.
Grasshopper Warbler, 29.
Great Spotted Woodpecker, 82.
Great Titmouse, 41.
Greenfinch, 59.
Green Sandpiper, 105.
Green Woodpecker, 81.
Gull, Common, 110.

Hawfinch, 58.
Hedge Sparrow, 20.
Heron, 118.
Herring Gull, 110.
Hobby, 3.
Hooded Crow, 67.

Hoopoe, 78.
House Martin, 73.
House Sparrow, 57.

Jackdaw, 69.
Jack Snipe, 102.
Jay, 71.

Kestrel, 5.
Kingfisher, 80.
Kittiwake Gull, 113.

Land Rail, 92.
Lapwing, 99.
Lesser black-backed Gull, 112.
Lesser Redpoll, 62.
Lesser spotted Woodpecker, 83.
Lesser Tern, 107.
Lesser Whitethroat, 31.
Linnet, 63.
Little Auk, 110.
Little Grebe, 117.
Long-eared Owl, 9.
Long-tailed Titmouse, 45.

Magpie, 70.
Marsh Titmouse, 44.
Meadow Pipit, 49.
Merlin, 4.
Mistletoe Thrush, 14.
Moor Hen, 95.

Night Heron, 119.
Nightingale, 22.
Nightjar, 76.
Nuthatch, 40.

Partridge, 89.
Peregrine Falcon, 2.
Pheasant, 88.
Pied Wagtail, 46.
Pochard, 125.

Quail, 91.

Red-backed Shrike, 12.
Redbreast, 21.
Red-legged Partridge, 90.
Redstart, 23.
Redwing, 16.
Reed Bunting, 51.
Reed Warbler, 27.
Ring Dove, 85.
Ringed Plover, 97.
Ring Ouzel, 19.
Roller, 79.
Rook, 68.

Sand Martin, 74.
Sandpiper, Common, 104.
Sedge Warbler, 28.
Short-eared Owl, 10.
Shoveller, 122.
Siskin, 61.
Skylark, 50.
Snipe, Common, 101.
Song Thrush, 15.
Sparrow Hawk, 6.
Spotted Crake, 93.
Spotted Flycatcher, 13.
Starling, 65.
Stock Dove, 86.

Stonechat, 24.
Storm Petrel, 115.
Swallow, 72.
Swift, 75.

Tawny Owl, 8.
Teal, 123.
Tree Creeper, 39.
Tree Pipit, 48.
Tree Sparrow, 56.
Turtle Dove, 87.

Water Rail, 94.
Wheatear, 26.
Whinchat, 29.
White-fronted Goose, 120
White-tailed Eagle, 1.
Whitethroat, 30.
Wigeon, 124.
Wild Duck, 121.
Willow Wren, 35.
Woodcock, 100.
Wood Wren, 34.
Wren, 38.
Wryneck, 84.

Yellow Hammer, 53.
Yellow Wagtail, 47.

INDEX TO THE PLANTS.

GENERIC NAMES.

(The numbers refer to the pages.)

Acer, 59.
Aceras, 105.
Achillea, 78.
Aconitum, 48.
Acorus, 111.
Adonis, 45.
Adoxa, 74.
Ægopodium, 71.
Æthusa, 72.
Agrimonia, 65.
Agropyron, 122.
Agrostemma, 54.
Agrostis, 117.
Aira, 118.
Ajuga, 96.
Alchemilla, 65.
Alisma, 112.
Allium, 108.
Alnus, 101.
Alopecurus, 117.
Anacharis, 104.
Anagallis, 86.
Anchusa, 87.
Anemone, 45.
Angelica, 73.
Anthemis, 78.
Anthoxanthum, 117.
Anthriscus, 72.
Anthyllis, 61.
Antirrhinum, 90.
Apargia, 83.
Apera, 118.
Apium, 70.
Aquilegia, 47.
Arabis, 50.
Arctium, 80.
Arenaria, 55.
Armoracia, 50.
Arrhenatherum, 119.
Artemisia, 79.
Arum, 111.
Arundo, 119.
Asparagus, 108.

Asperula, 75.
Aspidium, 124.
Asplenium, 123
Athyrium, 123.
Atriplex, 98.
Atropa, 89.
Avena, 118.

Ballota, 98.
Barbarea, 50.
Bartsia, 92.
Bellis, 77.
Berberis, 48.
Betula, 101.
Bidens, 78.
Blackstonia, 86.
Blysmus, 114.
Borago, 87.
Brachypodium, 122.
Brassica, 51.
Briza, 120.
Bromus, 121.
Bryonia, 70.
Bunium, 71.
Bupleurum, 70.
Butomus, 112.

Calamagrostis, 117.
Calamintha, 94.
Callitriche, 68.
Calluna, 84.
Caltha, 47.
Calystegia, 88.
Campanula, 84.
Capsella, 51.
Cardamine, 50.
Carduus, 80.
Carex, 114.
Carpinus, 101.
Castanea, 101.
Catabrosa, 119.
Caucalis, 73.
Centaurea, 81.

Centranthus, 76.
Cephalanthera, 105.
Cerastium, 54.
Ceratophyllum, 103.
Ceterach, 123.
Chærophyllum, 71.
Chara, 125.
Cheiranthus, 49.
Chelidonium, 49.
Chenopodium, 97.
Chlora, 86.
Chrysanthemum, 79.
Chrysosplenium, 67.
Cichorium, 82.
Circæa, 69.
Clematis, 45.
Cnicus, 81.
Cochlearia, 50.
Colchicum, 109.
Comarum, 65.
Conium, 70.
Conopodium, 71.
Convallaria, 108.
Convolvulus, 88.
Cornus, 74.
Coronopus, 51.
Corydalis, 49.
Corylus, 101.
Corynephorus, 118.
Cratægus, 67.
Crepis, 82.
Crocus, 107.
Cuscuta, 89.
Cynoglossum, 87.
Cynosurus, 119.
Cystopteris, 124.
Cytisus, 59.

Dactylis, 120.
Daphne, 99.
Daucus, 73.
Delphinium, 47.
Deschampsia, 118.

INDEX TO THE PLANTS.

Dianthus, 53.
Digitaria, 116.
Digraphis, 116.
Diplotaxis, 51.
Dipsacus, 76.
Doronicum, 80.
Draba, 50.

Echium, 88.
Eleocharis, 113.
Elodea, 104.
Epilobium, 69.
Epipactis, 105.
Equisetum, 125.
Eranthis, 47.
Erigeron, 77.
Eriophorum, 114.
Erodium, 58.
Erophila, 50.
Erysimum, 51.
Erythræa, 87.
Euonymus, 58.
Eupatorium, 77.
Euphorbia, 100.
Euphrasia, 92.

Fagus, 101.
Festuca, 120.
Filago, 77.
Fœniculum, 72.
Fragaria, 65.
Fraxinus, 86.
Fritillaria, 109.
Fumaria, 49.

Galanthus, 107.
Galeopsis, 95.
Galium, 75.
Genista, 59.
Geranium, 57.
Geum, 64.
Glyceria, 120.
Gnaphalium, 78.
Gymnadenia, 106.

Habenaria, 106.
Hedera, 73.
Helleborus, 47.
Helminthia, 82.
Helosciadium, 70.
Heracleum, 73.
Hesperis, 50.
Hieracium, 82.
Hippocrepis, 62.
Hippuris, 68.
Holcus, 118.
Holosteum, 54.
Hordeum, 122.

Hottonia, 85.
Humulus, 100.
Hyacinthus, 108.
Hydrocharis, 104.
Hydrocotyle, 70.
Hyoscyamus, 89.
Hypericum, 56.
Hypochæris, 82.
Hypopitys, 85.

Ilex, 58.
Inula, 78.
Iris, 106.

Jasione, 84.
Juncus, 109.

Kœleria, 119.

Lactuca, 83.
Lamium, 96.
Lapsana, 82.
Lastræa, 124.
Lathyrus, 62.
Lemna, 111.
Leontodon, 83.
Leonurus, 96.
Lepidium, 52.
Lepigonum, 56.
Ligustrum, 86.
Lilium, 109.
Linaria, 90.
Linum, 57.
Listera, 104.
Lithospermum, 88.
Lolium, 122.
Lonicera, 74.
Lotus, 61.
Luzula, 110.
Lychnis, 54.
Lycium, 89.
Lycopsis, 87.
Lycopus, 94.
Lysimachia, 85.
Lythrum, 69.

Malva, 56.
Marrubium, 95.
Matricaria, 79.
Medicago, 60.
Melampyrum, 92.
Melica, 119.
Melilotus, 60.
Mentha, 93.
Menyanthes, 87.
Mercurialis, 100.
Milium, 117.
Mœnchia, 54.

Monotropa, 85.
Muscari, 108.
Myosotis, 87.
Myosurus, 45.
Myriophyllum, 68.

Narcissus, 107.
Nasturtium, 49.
Neottia, 104.
Nepeta, 95.
Nephrodium, 124.
Nuphar, 48.
Nymphæa, 48.

Œnanthe, 72.
Œnothera, 69.
Onobrychis, 62.
Ononis, 59.
Onopordon, 81.
Ophioglossum, 124.
Ophrys, 106.
Orchis, 105.
Origanum, 94.
Ornithogalum, 108.
Ornithopus, 61.
Orobanche, 92.
Oxalis, 58.

Panicum, 116.
Papaver, 48.
Parietaria, 101.
Paris, 109.
Parnassia, 67.
Pastinaca, 73.
Pedicularis, 92.
Peplis, 69.
Petasites, 79.
Peucedanum, 73.
Phalaris, 116.
Phleum, 117.
Phragmites, 119.
Picris, 82.
Pimpinella, 71.
Pinus, 103.
Plantago, 97.
Poa, 120.
Polygala, 53.
Polygonum, 98.
Polypodium, 124.
Polystichum, 124.
Populus, 102.
Potamogeton, 112.
Potentilla, 65.
Poterium, 66.
Primula, 85.
Prunella, 95.
Prunus, 63.

GENERIC NAMES.

Pteris, 123.
Pulicaria, 78.
Pyrus, 66.

Quercus, 101.

Ranunculus, 46
Raphanus, 52.
Reseda, 52.
Rhamnus, 59.
Rhinanthus, 92.
Ribes, 67.
Rosa, 66.
Rubus, 63.
Rumex, 99.
Ruscus, 107.

Sagina, 55.
Sagittaria, 112.
Salix, 102.
Salvia, 94.
Sambucus, 74.
Samolus, 86.
Sanicula, 70.
Saponaria, 53.
Sarothamnus, 59.
Saxifraga, 67.
Scabiosa, 77.
Scandix, 71.
Scilla, 108.
Scirpus, 114.
Scleranthus, 97.
Sclerochloa, 120.
Scolopendrium, 124.
Scrophularia, 90.
Scutellaria, 95.

Sedum, 68.
Sempervivum, 68.
Senebiera, 51.
Senecio, 80.
Serrafalcus, 121.
Setaria, 116.
Sherardia, 76.
Sieglingia, 119.
Silaus, 73.
Silene, 53.
Silybum, 81.
Sinapis, 51.
Sison, 71.
Sisymbrium, 50.
Sium, 71.
Smyrnium, 70.
Solanum, 89.
Solidago, 77.
Sonchus, 83.
Sparganium, 111.
Specularia, 84.
Spergula, 55.
Spergularia, 56.
Spiræa, 63.
Spiranthes, 105.
Stachys, 95.
Stellaria, 55.
Stratiotes, 104.
Symphytum, 87.

Tamus, 107.
Tanacetum, 79.
Taraxacum, 83.
Taxus, 103.
Teesdalia, 52.
Teucrium, 96.

Thalictrum, 45.
Thlaspi, 52.
Thrincia, 83.
Thymus, 94.
Tilia, 57.
Tillæa, 67.
Torilis, 73.
Tragopogon, 84.
Trifolium, 60.
Triglochin, 112.
Triodia, 119.
Trisetum, 118.
Triticum, 122.
Tulipa, 109.
Turritis, 50.
Tussilago, 79.
Typha, 110.

Ulex, 59.
Ulmus, 100.
Urtica, 100.
Utricularia, 93.

Valeriana, 76.
Valerianella, 76.
Verbascum, 90.
Verbena, 93.
Veronica, 91.
Viburnum, 74.
Vicia, 62.
Vinca, 86.
Viola, 52.
Viscum, 99.

Zannichellia, 113.

INDEX TO THE PLANTS.

ENGLISH NAMES.

(The numbers refer to the pages.)

Aconite, 48.
Adder's Tongue, 124.
Agrimony, 65.
Alder, 101.
Ale Hoof, 95.
Alexanders, 70.
Alkanet, 87.
Anemone, 45.
Angelica, 73.
Apple, 66.
Archangel, 96.
Arrow-grass, 112.
Arrow-head, 112.
Ash, 86.
Asparagus, 108.
Aspen, 102.
Avens, 64.

Barberry, 48.
Barley, 122.
Bartsia, 92.
Basil, 94.
Bedstraw, 75.
Beech, 101.
Bell-flower, 84.
Bent Grass, 117.
Betony, Wood, 95.
Betony, Water, 90.
Bind-weed, 88.
Birch, 101.
Bird Cherry, 63.
Bird's-foot, 61.
Bird's-nest, 85.
Bistort, 98.
Bittersweet, 89.
Blackberry, 63.
Black Bryony, 107.
Black Thorn, 63.
Bladder Fern, 124.
Bladderwort, 93.
Bluebell, 108.
Blysmus, 114.
Borage, 87.

Box Thorn, 89.
Brake, 123.
Bramble, 63.
Briar, 66.
Bristle Grass, 116.
Brome Grass, 121.
Brooklime, 91.
Brookweed, 86.
Broom, 59.
Broom-rape, 92.
Bryony, 70, 107.
Buckbean, 87.
Buckler Fern, 124.
Buckthorn, 59.
Buckwheat, 98.
Bugle, 96.
Bugloss, 88.
Bullace, 63.
Bullrush, 110, 114.
Burdock, 80.
Bur-Marigold, 78.
Burnet, 66.
Burnet Saxifrage, 71.
Bur-reed, 111.
Butcher's Broom, 107.
Butter-bur, 79.
Buttercup, 46.

Calamint, 94.
Campion, 54.
Canary Grass, 116.
Carrot, 73.
Catch-fly, 53.
Cat Mint, 95.
Cat's-ear, 82.
Cat's-tail, 110.
Celandine, 47, 49.
Centaury, 81.
Chamomile, 78.
Chara, 125.
Charlock, 51.
Cherry, 63.
Chervil, 71, 72.

Chick-weed, 55.
Chicory, 82.
Cinque-foil, 65.
Clary, 94.
Cleavers, 75.
Clover, 60.
Club-rush, 114.
Cock's-foot Grass, 120.
Colt's-foot, 79.
Columbine, 47.
Comfrey, 87.
Convolvulus, 88.
Corncockle, 54.
Corn Marigold, 79.
Cotton Grass, 114.
Couch Grass, 122.
Cow Parsnip, 73.
Cowslip, 85.
Cow-wheat, 92.
Crane's-bill, 58.
Creeping Jenny, 85.
Crocus, 107.
Crosswort, 75.
Crowfoot, 46.
Cuckoo Flower, 50, 54, 105.
Cuckoo-pint, 111.
Cudweed, 77, 78.
Currant, 67.

Daffodil, 107.
Daisy, 77.
Dame's-Gilliflower, 50.
Dandelion, 83.
Danewort, 74.
Darnel, 122.
Dead Nettle, 96.
Deadly Nightshade, 89.
Dewberry, 64.
Dock, 80.
Dodder, 89.
Dog Rose, 66.
Dog's-tail Grass, 119.

INDEX TO THE PLANTS.

Dog Wood, 74.
Drauk, 121.
Dropwort, 72.
Duckweed, 111.
Dyer's Green-weed, 59
Dyer's Weld, 52.

Earth-nut, 71.
Elder, 74.
Elm, 100.
Enchanter's Nightshade, 69.
Evening Primrose, 69.
Eye-bright, 92.

Fennel, 72.
Ferns, 123.
Fescue Grass, 120.
Feverfew, 79.
Field Madder, 76.
Figwort, 90.
Finger Grass, 116.
Fir, 103.
Flag, 106.
Flax, 57.
Flea-bane, 77, 78.
Flixweed, 51.
Flowering Rush, 112.
Fluellin, 90.
Fool's Parsley, 72.
Forget-me-not, 87.
Fox-tail Grass, 117.
Fritillary, 109.
Frog-bit, 104.
Frog Orchis, 106.
Fumitory, 49.
Furze, 59.

Garlic, 108.
Garlic Mustard, 51.
Gipsywort, 94.
Goat's Beard, 84.
Golden Rod, 77.
Golden Saxifrage, 67.
Goldilocks, 46.
Good King Harry, 97.
Gooseberry, 67.
Goose-foot, 97.
Goose-grass, 75.
Gorse, 59.
Gout-weed, 71.
Grape Hyacinth, 108.
Grass of Parnassus, 67.
Green-weed, 59.
Gromwell, 88.
Ground Ivy, 95.
Groundsel, 80.
Guelder Rose, 74.

Hair Grass, 118.
Harebell, 84.
Hare's-ear, 70.
Hart's-tongue, 124.
Hawk's-beard, 82.
Hawkbit, 83.
Hawkweed, 82.
Hawthorn, 67.
Hazel, 101.
Heart's-ease, 53.
Heath Grass, 119.
Hedge-mustard, 51.
Hedge Parsley, 73.
Hellebore, 47.
Helleborine, 105.
Hemlock, 70.
Hemp Agrimony, 77.
Hemp Nettle, 95.
Henbane, 89.
Herb Paris, 109.
Herb Robert, 58.
Holly, 58.
Honeysuckle, 74.
Hop, 100.
Horehound, 95, 98.
Hornbeam, 101.
Horned Pondweed, 113
Hornwort, 103.
Horse Radish, 50.
Horse-tail, 125.
Hound's-tongue, 87.
House-leek, 68.
Hyacinth, 108.
Hyssop, 94.

Iris, 106.
Ivy, 73.
Ivy, Ground, 95.

Kidney Vetch, 61.
Knapweed, 81.
Knawel, 97.
Knot-grass, 98.

Lady Fern, 123.
Lady's-mantle, 65.
Lady's-slippers, 61.
Lady's-smock, 50.
Lady's-tresses, 105.
Lamb's Lettuce, 76.
Larkspur, 47.
Leopard's-bane, 80.
Lettuce, 83.
Lily of the Valley, 108.
Lily, Turk's-cap, 109.
Lily, Water, 48.
Lime, 57.
Ling, 84.

Loose-strife, 85.
Loose-strife, Purple, 69.
Lords and Ladies, 111.
Lousewort, 92.
Lucerne, 60.

Madder, 76.
Male Fern, 124.
Mallow, 56.
Man Orchis, 105.
Maple, 59.
Mare's-tail, 68.
Marigold, 47, 79.
Marjoram, 94.
Marsh Marigold, 47.
May, 67.
Mayweed, 78, 79.
Meadow Grass, 120.
Meadow Rue, 45.
Meadow-sweet, 63.
Medick, 60.
Melic Grass, 119.
Melilot, 60.
Mercury, 100.
Mezereon, 99.
Mignonette, 52.
Milfoil, 68, 78.
Milk Thistle, 81.
Milkwort, 53.
Millet Grass, 117.
Mint, 93.
Mistletoe, 99.
Moneywort, 85.
Monk's-hood, 48.
Moschatel, 74.
Motherwort, 96.
Mouse-tail, 45.
Mugwort, 79.
Mullein, 90.
Mustard, 51.

Narcissus, 107.
Nettle, 100.
Nettle, Dead, 96.
Nightshade, 89.
Nipplewort, 82.
Nonsuch, 60.

Oak, 101.
Oat Grass, 118, 119.
Orache, 98.
Orchis, 105, 106.
Orpine, 68.
Osier, 102.
Ox-eye, 79.
Ox-lip, 85.
Ox-tongue, 82.

Pansy, 53.
Parnassus, G. of, 67.
Parsley, 72, 73.
Parsnip, 70, 71, 73.
Pear, 66.
Pearlwort, 55.
Pellitory, 101.
Penny-cress, 52.
Penny-royal, 94.
Pennywort, 70.
Peppermint, 93.
Pepperwort, 52.
Periwinkle, 86.
Persicaria, 98.
Pheasant's-eye, 45.
Pimpernel, 86.
Pink, 53.
Plaintain, 97.
Plaintain, Water, 112.
Ploughman's Spikenard, 78.
Plum, 63.
Polypody, 124.
Pondweed, 112.
Poor Man's Weatherglass, 86.
Poplar, 102.
Poppy, 48.
Primrose, 85.
Privet, 86.
Purslane, 69.

Quaking Grass, 120.

Radish, 52.
Ragged Robin, 54.
Ragwort, 80.
Ramsons, 108.
Rape, 51.
Raspberry, 63.
Rattle, 92.
Red-berried Bryony, 70.
Reed, 119.
Reed-mace, 110.
Rest-harrow, 59.
Robin Hood, 54.
Rocket, 51.
Rose, 66.
Rue, Meadow, 45.
Rush, 109.
Rush, Wood, 110.
Rye Grass, 122.

Saffron, 109.
Sage, 96.
Sainfoin, 62.
Sallow, 102.
Salad-Burnet, 66.
Sandwort, 55.
Sanicle, 70.
Saxifrage, 67.
Scabious, 77.
Sedge, 114.
Self-heal, 95.
Service Tree, 66.
Sheep's-bit, 84.
Shepherd's-needle, 71.
Shepherds'-purse, 51.
Shield Fern, 124.
Silver Weed, 65.
Skull-cap, 95.
Sloe, 63.
Small Reed, 117.
Snapdragon, 90.
Snowdrop, 107.
Soapwort, 53.
Soft Grass, 118.
Sorrel, 58, 99.
Sorrel, Wood, 58.
Sow-Thistle, 83.
Spearwort, 46.
Speedwell, 91.
Spike-Rush, 113.
Spindle Tree, 58.
Spleenwort, 123.
Spurge, 100.
Spurge Laurel, 99.
Spurrey, 55, 56.
St. John's-wort, 56.
Star of Bethlehem, 108.
Starwort, 68.
Stitchwort, 55.
Stone-crop, 68.
Stonewort, 71.
Stork's-bill, 58.
Strawberry, 65.
Succory, 82.
Sulphurwort, 73.
Sweet Briar, 66.
Sweet Flag, 111.
Sycamore, 59.

Tansy, 79.
Tare, 62.
Teasel, 76.
Teesdalia, 52.
Thale Cress, 50.

Thistle, 80, 81.
Thyme, 94.
Tillæa, 67.
Timothy Grass, 117.
Tinker-tailor Grass, 122.
Toad-flax, 90.
Tormentil, 65.
Tower Mustard, 50.
Travellers' Joy, 45.
Trefoil, 60.
Tulip, 109.
Turnip, 51.
Tway-blade, 104.

Valerian, 76.
Venus' Comb, 71.
Venus' Looking-glass, 84.
Vernal Grass, 117.
Vervain, 93.
Vetch, 62.
Vetchling, 62.
Violet, 52.
Violet, Water, 85.
Viper's Bugloss, 88.

Wall Flower, 49.
Wart-cress, 51.
Water Cress, 49.
Water Dropwort, 72.
Water Lily, 48.
Water Milfoil, 68.
Water Parsnip, 70, 71.
Water Pepper, 98.
Water Plaintain, 112.
Water-Soldier, 104.
Water Thyme, 104.
Water Violet, 85.
Wayfaring Tree, 74.
Weld, 52.
White Beam, 66.
Whitlow Grass, 50.
Whorl Grass, 119.
Willow, 102.
Willow-herb, 69.
Woodruff, 75.
Wood Rush, 110.
Wood Sage, 96.
Wood Sorrel, 58.
Woundwort, 95.

Yarrow, 78.
Yellow-cress, 49.
Yew, 103.

www.ingramcontent.com/pod-product-compliance
Lightning Source LLC
Chambersburg PA
CBHW030319170426
43202CB00009B/1074